# Global Magmatic Oxides:
# Diverse & Puzzling

# Global Magmatic Oxides: Diverse & Puzzling

Olsen

Horace, my friend. This crazy adventure wouldn't
have been the same without you by my side

# Table of Contents

# Chapter 1  Introduction

Magmatic Fe-Ti-V-P deposits (the term magmatic oxide deposits will be used from here on) located around the world potentially contain economic resources of Ti, V, or P. They are normally hosted within mafic/ultramafic rocks within layered intrusions (e.g., Bushveld Complex, Republic of South Africa (Wager and Brown 1968; Willemse 1969; von Gruenewaldt 1973), Sept Iles, Canada (Tollari et al. 2008)) or hosted within anorthosite or leucogabbro units of Proterozoic anorthosite-mangerite-charnockite-granite (AMCG) complexes (e.g., Lac Tio, Canada; Tellnes, Norway; or Damiao, China) (Charlier et al. 2015). In layered intrusions, magmatic oxide deposits consist of stratiform layers of magnetite $\pm$ ilmenite $\pm$ apatite as late stage cumulates (Charlier et al. 2015) but, in anorthosite complexes, display the form of massive, tabular, or lenticular bodies in the host anorthosite (Hébert et al. 2005; Charlier et al. 2015). Each mineralization style can display any of these mineral assemblages; 1) (gabbro-) noritic ilmenite ore $\pm$ magnetite $\pm$ apatite, 2) Ti-magnetite $\pm$ apatite ore, 3) nelsonite, a rock composed of roughly two thirds Fe-oxides and one third apatite (Philpotts 1967; Kolker 1982), or 4) rutile-ilmenite ore (Charlier et al. 2015). Though these deposits have been studied in abundance, questions still remain regarding the genesis of these deposits and the method/s of concentrating ore minerals. The following models have been proposed; 1) fractional crystallization followed by either gravity-driven or flow-driven separation from silicates, 2) separation of an immiscible Fe-Ti $\pm$ P-rich magma from a silicate magma in a way similar to that of their magmatic Ni-Cu counterparts, and 3) solid-state remobilization (Charlier et al. 2015).

The Grenville Province, primarily located in Canada from southeastern Labrador through Quebec to southwestern Ontario, contains several large AMCG complexes which host several occurrences of magmatic Fe-Ti-V-P mineralization (Corriveau et al. 2007). The largest of those belong to the Geon-11 magmatic episode where anorthositic magmatism was emplaced from 1170-1110 Ma (Rivers et al 2012). Mineralization associated with the anorthosite units in these AMCG suites can either be composed of Ti-magnetite, ilmenite, ± apatite or (hemo-) ilmenite ± apatite ± magnetite (Corriveau et al. 2007). Varying mineralogy in these deposits are possibly due to 1) magma composition, 2) crustal contamination, or 3) $fO_2$ (Woodruff et al. 2013).

The Lac St. Jean anorthosite suite is one of the largest (20 000 km²) Mesoproterozoic AMCG complexes in the world and is located within the Grenville Province near Saguenay, Québec, Canada. Due to its large volume, accessible location, and the abundance of mineralization along the margins of the pluton it appears to be an ideal area to study the genesis of magmatic Fe-Ti-V-P mineralization. The first regional-scale study on Fe-Ti-V-P mineralization within the Lac St. Jean anorthosite suite was completed by Hébert et al. (2005) who examined mineralization along the northern, eastern, and southern margins. While concluding that Fe-Ti-P mineralization (Fe-oxides + apatite), which may contain hemo-ilmenite, is associated with andesine-type anorthosite whereas Fe-Ti-V mineralization (Ti-magnetite ± ilmenite) is isolated to labradorite-type anorthosites, Hébert et al. (2005) also organized mineralization styles within the northern margin of the Lac St. Jean anorthosite suite into three categories; 1) Ti-Fe deposits containing Ti-magnetite (± ilmenite) layers, dykes, and breccias, 2) dyke-like, concordant lenses of nelsonite + Ti-Fe-P-bearing mafic/ultramafic lithologies, and 3) apatite-rich (5-8% $P_2O_5$) andesine anorthosites, leuconorites and norites. The main difference between the northern and eastern margins of the Lac St. Jean anorthosite suite is the presence of dunite and peridotite units and the abundance of oxide apatite gabbro/norite (OAGN) units respectively (Hébert et al. 2005). However, the processes that control this regional-scale variation in Fe-oxide and apatite mineralogy of Fe-Ti-V-P mineralization is not understood.

Results presented in this study will be the first regional-scale study on several magmatic Fe-Ti-V-P deposits of the Lac St. Jean anorthosite suite involving the combination of detailed petrography and mineral chemistry of Fe-oxides and apatite using LA-ICP-MS. This project aims to add insight on the genesis of these deposits based on characterising the mineralogical and geochemical variation of Fe-Ti-V-P mineralization at the regional- and deposit-scale. Advances in laser ablation (LA)-ICP-MS now permit accurate analysis of trace elements, which normally would have been below detection limit using electron microprobe and XRF on mineral separates. Previous studies using trace element compositions of Fe-oxides and/or apatite have proven quite useful in constraining petrogenetic models for the world class magmatic oxide deposits associated with layered intrusions, such as Sept Iles (Tollari et al. 2008; Namur et al. 2010) and Proterozoic anorthosite complexes, such as Lac Tio, Québec (Charlier et al. 2011), Tellnes, Norway (Charlier et al. 2007), and Damiao, China (He et al. 2016). Other workings have also indicated that the composition of Fe-oxides can be modified during sub-solidus re-equilibration between Fe-oxide phases (i.e. magnetite, ilmenite, and Al-spinel, Frost 1991; Pang et al 2007) and Fe-Mg silicate minerals (Morse 1980; Frost et al. 1988; Frost 1991; Frost and Lindsley 1992; Pang et al. 2007; Charlier et al. 2015).

Despite previous research, several questions remain surrounding magmatic Fe-Ti-V-P deposits hosted in anorthosite complexes, involving the relationship with their host rocks, regional-, deposit- scale variations, and petrogenetic models, which this study will address.

# Chapter 2   Geological Setting

The Grenville Province is an approximately 5000 km long by 600-800 km wide orogen which is mostly underlain by gneiss complexes with polyphase ductile deformation and extensive partial melting (Rivers et al. 2012). The age of Grenvillian metamorphism and temperature and pressure conditions of peak metamorphism indicates two tectonic divisions of the Grenville province; 1) The Parautochthonous belt (Rivers and Chown 1986; Wardle et al. 1986; Rivers et al. 1989) and 2) Allochthonous belts (Ludden and Hynes 2000; Rivers et al. 2002; Rivers 2008). Parautochthonous belt refers to belts/terranes/domains which lie adjacent to the Grenville front (Fig. 2.1) and not transported far from their basement whereas Allochthonous belt refers to belts/terranes/domains structurally above the Allochthon boundary thrust (ABT; Fig 2.1), which have been substantially displaced during Grenvillian orogenesis (Rivers et al. 2012). The allochthonous belts have been subdivided into high-pressure (HP) and medium to low-pressure (MP-LP) belts based on eclogite to upper-granulite and upper-amphibolite to granulite facies metamorphic assemblages respectively (Rivers et al. 2012). High-pressure metamorphic assemblages (HP-belt) in the allochthonous belt are dispersed locally along the ABT (Rivers et al. 2012) whereas MP-LP assemblages (MP-LP belt) extend the full length of the Grenville province (Rivers et al. 2008).

## 2.1   Tectonics and Magmatism

The Grenville province was formed through several Andean-style orogenic events on the SE-margin of Laurentia (Rivers et al. 2012) followed by a continent-continent collisional event between Laurentia and Amazonia (Tohver et al. 2006). Andean-style orogenic activity and crustal growth on SE-Laurentia is marked by the; 1) Labradorian

(1.71-1.60 Ga; Gower and Krogh 2002; Gower et al 2008), 2) Pinwarian (1.52-1.46 Ga; Tucker and Gower 1990), and 3) Elsonian and Elzevirian (1.40-1.20 and 1.25-1.18 Ga respectively; Carr et al. 2000) tectonic events. Continent-continent collision is marked by the Grenvillian orogeny (1100-980 Ma: Rivers et al. 2012). However, a possible short-lived period of extension shortly after crustal growth and prior to continent-continent collision is indicated by the intrusion of large, voluminous AMCG suites associated with the Geon-11 (1180-1110 Ma; Figs. 2.1 and 2.2) time period (Rivers et al. 2012).

Geon-11 AMCG magmatism is located within the allochtonous MP-LP belt of the Grenville Province (Rivers et al. 1989) and is oriented along two NS-trending corridors dividing the allochtonous MP-LP belt into three parts; Western, Central, and Eastern (Indares and Moukhsil 2013). However, magmatism is most abundant and diverse within the allochtonous MP belt of the central Grenville province (Figs. 2.1 and 2.2; Augland et al. 2015; Dunning and Indares 2010; Hébert et al. 2009; Moukhsil et al. 2009; Owens and Tomascack 2002) which is the location of the Lac St. Jean anorthosite suite. The proposed tectonic setting for Geon 11 AMCG magmatism remains only partially constrained but, its large volume and plagioclase compositions of labradorite-type (approximately An 60), indicates mantle melting initiated by back-arc rifting, due to slab rollback or ridge subduction, during the final closure of a Mesoproterozoic ocean at Southeast Laurentia (Rivers et al. 2012).

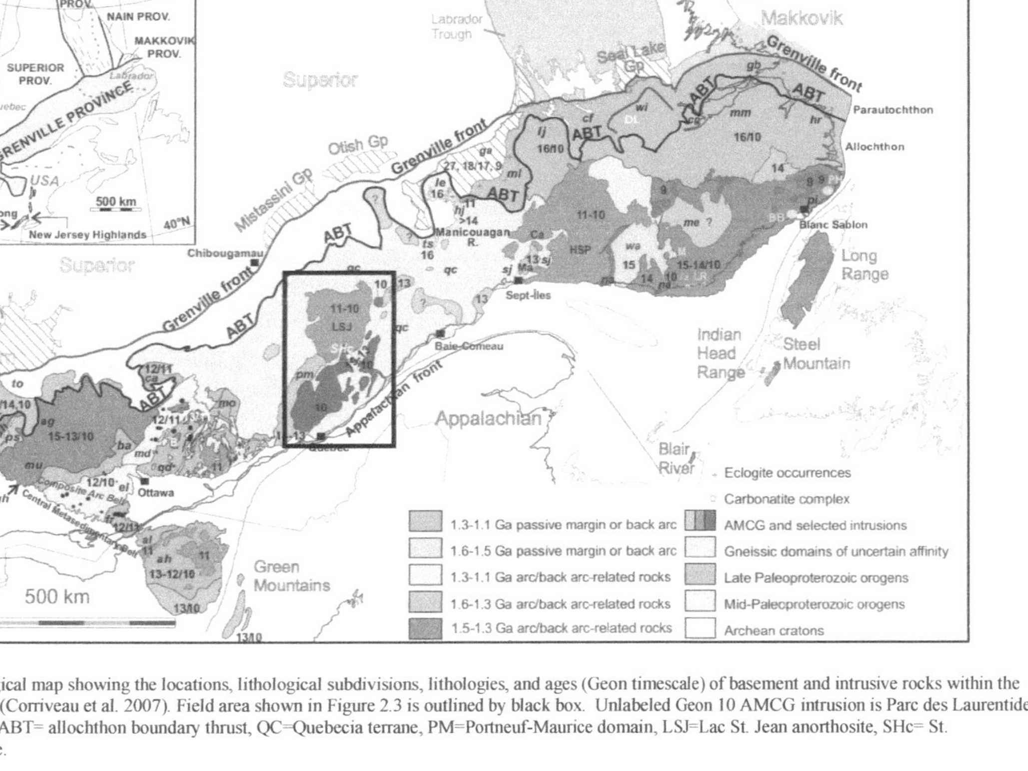


**Figure 2.1:** Geological map showing the locations, lithological subdivisions, lithologies, and ages (Geon timescale) of basement and intrusive rocks within the Grenville Province (Corriveau et al. 2007). Field area shown in Figure 2.3 is outlined by black box. Unlabeled Geon 10 AMCG intrusion is Parc des Laurentides intrusive complex. ABT= allochthon boundary thrust, QC=Quebecia terrane, PM=Portneuf-Maurice domain, LSJ=Lac St. Jean anorthosite, SHc= St. Honoré=carbonatite.

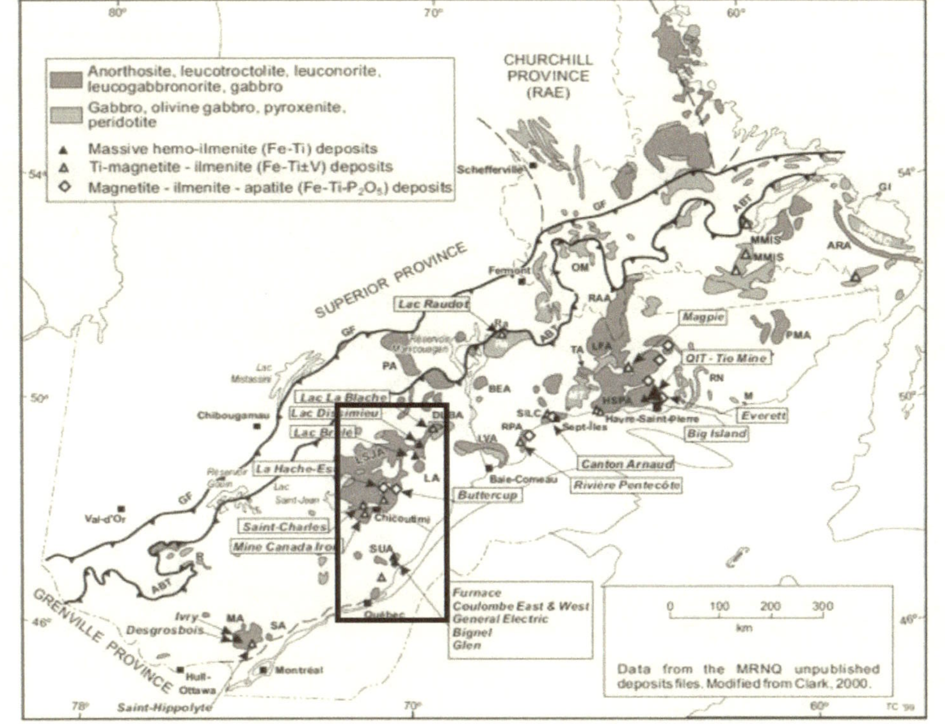


**Figure 2.2:** Geological map showing the names and locations of only the anorthositic and mafic/ultramafic intrusions within the Grenville Province and their associated Fe-oxide mineralization (Corriveau et al. 2007). Field area shown in figure 2.3 is outlined by black box. LSJA=Lac St. Jean anorthosite.

## 2.2 Local Geology

The basement rocks in the study area of the Lac St. Jean anorthosite suite, as summarized by Higgins et al. (2002), are a complex assemblage of plutonic rocks (now orthogneiss) intruded into paragneiss complexes, later crosscut by parallel shear zones (Fig. 2.3) that formed on the continental margin of Laurentia (1.5-1.3Ga), which has been coined Quebecia for the Central part of the Grenville (Dickin and Higgins 1992; Dickin 2000, Dickin et al. 2010; Vautour 2015). The basement rocks consist of 1) paragneiss and mafic-intermediate orthogneiss of the 1506 ± 13 Ma (Hébert and van Breeman 2004) Saguenay gneiss complex, representing supracrustal rocks and plutonic rocks respectively; 2) tonalitic to dioritic orthogneiss of the 1434 +64/-28 Ma (Gobeil et al. 2002) Hulot complex; and 3) widespread felsic with some charnockitic gneisses of the 1393-1383 Ma (Hébert and van Breeman 2004) Cap à l'Est gneiss complex (Hébert et al. 2005).

AMCG magmatism within the Lac St. Jean area commenced with the 1327 ± 16 Ma (Gobeil et al. 2002) De La Blanche anorthosite suite (labradorite-type) composed of mafic plutonic rocks: a core of anorthosite, leucotroctolite, and leuconoritic with a minor rim of gabbronorite and felsic rocks (Hébert et al. 2005). The De la Blanche anorthosite suite was once considered an extension of the Lac St. Jean anorthosite suite but, geochronological evidence from Gobeil et al. (2002) refutes this claim (Fig. 2.3).

The Lac St. Jean anorthosite suite occupies the central portion of the Lac St. Jean geological area (Fig.2.3) and has been mapped in detail by the Quebec Ministry of Energy and Natural Resources (Hébert and Lacoste 1998*a*, 1998*b*, 1998*c*, 1998*d*, 1998*e*; Hébert 1999, 2000, 2001, 2002, 2003; Hébert et al. 1999*a*, 1999*b*, 1999*c*, 2003; Hébert and Beaumier 2000*a*, 2000*b*; Hébert and van Breemen 2001; Hébert and Cadieux 2003).

Ninety five percent of this suite's volume consists of anorthosite and leuconorite while the other 5% consists of troctolite, norite, olivine gabbro, gabbro, pyroxenite, peridotite, dunite, ferrodiorite, nelsonite, magnetitite, and rare charnockite and mangerite (Hébert et al. 2005). Labradorite-type plagioclase dominates but, in the northern and eastern sections of the Lac St. Jean anorthosite suite, andesine-type is present (Hébert et al. 2005). Emplacement of the Lac St. Jean anorthosite suite commenced with 1160-1140 Ma anorthosite units (Higgins and van Breeman 1992; Higgins et al. 2002) and ended with minor 1155-1135 Ma charnockite and mangerite units (Hébert and van Breeman 2004). Fe-Ti-V-P mineralization is mainly concentrated in the northern portion of the Lac St. Jean anorthosite suite, in particular the region of Lac à Paul (Hébert et al. 2005; Fredette 2006) but, the eastern and southern portions also contain significant mineralization (Hébert et al. 2005) (Fig. 2.4). Ferrodiorite dykes in the region of Lac à Paul are considered to be the parental magma of Fe-Ti-V-P mineralization (Fredette, 2006). Ferrodiorites possibly represent the residual magma from a plagioclase-rich crystal mush which settles along the margin of anorthosite plutons along with mafic/ultramafic lithologies (Vander Auwera et al. 2006).

Several smaller AMCG-type plutons were later emplaced both into the Lac St. Jean anorthosite suite itself and the surrounding area, including many anorthositic and mafic intrusions belonging to the CRUML (Chateau Richer, St. Urbain, Lac Chaudiere, Lac a Jack, Mattawa, and Labrieville) belt massifs and other associated felsic intrusions (Fig. 2.3) (Hébert et al. 2005). Some of the anorthositic intrusions are host to Fe-Ti-V-P mineralization composed of hemo-ilmenite + apatite bodies within gabbronoritic rocks, such as the Lac à l'Orignal Fe-Ti-P occurrence (Glen Eagle Resources Inc. 2012) this study) in the Vanel anorthosite of the Pipmuacan suite (Hébert et al. 2005; 2009). In order of age, they include 1080-1059 Ma AMCG magmatism of the Pipmuacan anorthosite suite (St. Urbain and Vanel), the 1016-1008 Ma Valin AMCG suite (Mattawa and Labrieville), granitic rocks of the 1028-1018 Ma Peribonka plutonic suite, granitic to dioritic rocks of the 988-985 Ma Grand Points alkalic plutonic suite, and other smaller unnamed anorthosite, mafic, and felsic plutons (Fig. 2.3) (Hébert et al. 2009). Anorthosites within these younger intrusions are predominantly andesine in composition (Dymek and Owens 1998; Dymek 2001).

## 2.3   Structural Geology

The first deformation event within the Lac St. Jean area is marked by three major NE oriented shear zones parallel to other shear zones within the Grenville Province (Fig. 2.3) (Higgins et al. 2002). The Chute-des-Passes shear zone crosscuts the northern-most units of the Lac St. Jean anorthosite suite and is defined by a gneiss zone which can be as wide as 1km (Hébert 2009). The Pipmuacan shear zone (PSZ) crosscuts the central part of the Lac St. Jean anorthosite suite and is traced from the northeastern extremity of the Pipmuacan reservoir to near Lac St. Jean (Hébert 1991). It is identified by solid-state deformation, mylonites, and faults oriented N050 to N60 (Higgins et al. 2002). The St. Fulgence shear zone follows the southeastern contact of the Lac St. Jean anorthosite suite (Fig. 2.3) but, it is unclear if this controlled or postdates emplacement (Higgins et al. 2002). Movement along this fault is similar to that along the Pipmuacan shear zone (Hébert and Lacoste 1998a).

The second set of shear zones within the Lac St. Jean area are oriented north to north-north east (Fig. 2.3) (Gervais 1993; Daigneault 1999). They are marked by a narrow zone of straight gneiss with near vertical orientations (Higgins et al. 2002). Rare structural indicators suggest that movement along the shear zone is sinistral with a thrust component. Deformation in the 1157 Ma megadykes potentially constrains the age of shear zone movement (Higgins et al. 2002).

The youngest deformation events within the area are unnamed northwest oriented faults marked by mylonites and injections of granite in places, meaning that it was active during late Grenvillian times (Fig. 2.3) (Higgins et al. 2002). No geochronology work has been performed on these faults. However, post-tectonic intrusions such as the Astra granite (1020 ± 2 Ma;) and the Venus de Milot intrusion (988 ± 3 Ma) helped guide determination of the age of these faults but, they have been reactivated later and crosscut late Grenvillian intrusions (Higgins et al. 2002).

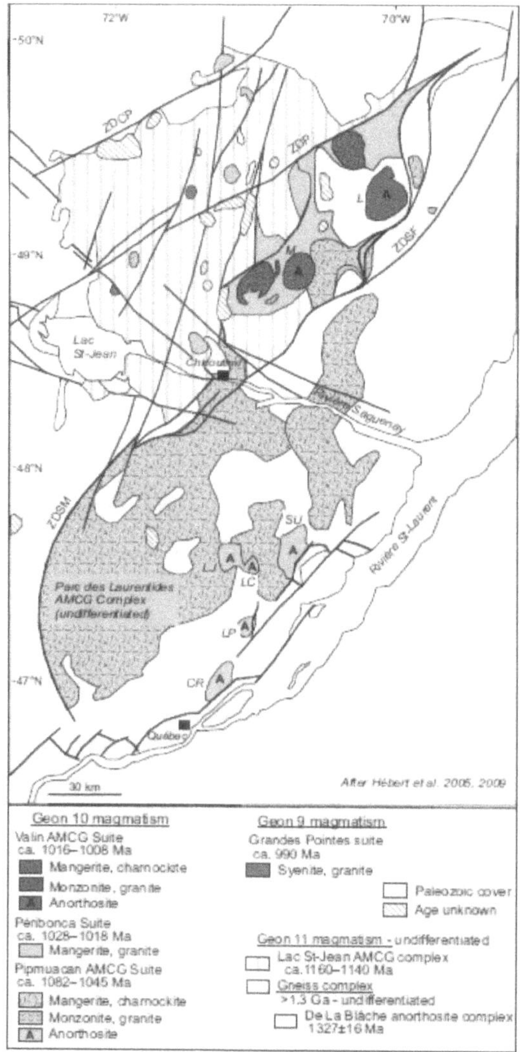


**Figure 2.3** Geological map of the Lac St. Jean area displaying and the 6 different ages of AMCG related magmatism intruded into basement gneiss complex (Rivers et al. 2012 modified from Hébert et al 2009). Abbreviations: ZDCP, ZDP, ZDSF, ZDSM represent Chute des Passes, Pipmuacan, St. Fulgence, and St. Maurice deformation zones; CR-Château Richer, L-Labriville, LC-Lac Chaudière, LJ-Lac à Jack, LP-Lac Piché, M-Mattawa, SU-St. Urbain.

13

2.4    Geology of oxide-apatite mineralization

In total, 6 deposits/occurrences of oxide-apatite mineralization, all of which are located near the margins of the Lac St. Jean anorthosite suite, were examined in this study (Fig. 2.4, Table 2.1). They were chosen to illustrate the variety of oxide mineralogy in the Lac St. Jean area. The mineralization is categorized into Fe-Ti-V rich (Buttercup and Lac Margane/Houlière-Nord; Fig. 2.4A) and Fe-Ti-P rich (St. Charles de Bourget, Lac Perron, Lac à Paul, and Lac à l'Orignal; Fig 2.4B) based on the absence or presence of apatite respectively. Some deposits contain both Fe-Ti-V and Fe-Ti-P rich mineralization (St. Charles de Bourget and Lac à l'Orignal) but, are classified as Fe-Ti-P for this study because they do contain significant apatite resources. On the outcrop-scale, the Fe-oxides $\pm$ apatite concentrations display two main habits; 1) concordant/discordant massive pods hosted within anorthosite (e.g., Buttercup, Lac Perron), in some cases shear-zone associated (e.g., St. Charles de Bourget) or 2) massive to semi-massive layers/foliations alternating with mafic/ultramafic lithologies (e.g., Lac Margane/Houlière and Lac à Paul). Each deposit also contains different magnetite to ilmenite proportions clearly visible on hand sample-scale.

Details of each deposit/occurrences are given in Chapters 3 (Fe-Ti-V) and 4 (Fe-Ti-P), but an overview is presented below to provide a summary of mineralization within the Lac St. Jean area.

2.4.1    Fe-Ti-V Deposits

The Buttercup deposit, located within the eastern margin of the Lac St. Jean anorthosite suite (Fig. 2.4A) perhaps contains the most significant V resources with 3.2 Mt at 49% $Fe_2O_3$, 19% $TiO_2$, and 0.67% $V_2O_5$ (Table 2.1) (Lalancette 2014).

Mineralization is composed of three small massive oxide lenses composed of coarse Ti-magnetite with rare ilmenite hosted within anorthosite (Lalancette 2014).

Located on the northern margin of the Lac St. Jean anorthosite are a group of Fe-Ti-V showings called Lac Margane and Lac Houlière-Nord (Fig. 2.4A). Since the Lac Margane and Lac Houlière-Nord deposits occur along strike of each other and have similar mineralization styles, they have been grouped as Lac Margane/Houlière-Nord for this project. Several whole rock analyses of mineralized samples from Tremblay (2016) has indicated 45.6% $Fe_2O_3$, 15.4% $TiO_2$, 0.30% $V_2O_5$ (Table 2.1); no tonnage data is presented. According to Tremblay (2016), the main mineralization is located near the contact between gabbronorite and leuconorite units and consists of decameter-scale bands of Fe-oxides alternating with gabbronorite or leuconorite.

### 2.4.2   Fe-Ti-P Deposits

The St. Charles de Bourget deposit (currently owned by Micrex) located on the southern margin of the Lac St. Jean anorthosite suite (Fig 2.4A, Table 2.1) is a much different deposit than those described earlier. It is composed of Ti-magnetite ± apatite pods, containing enclaves of gabbro-diorite, up to 100s of meters wide scattered randomly within anorthosite or anorthositic rocks with variable mafic minerals (IOS Services Géoscientifiques Inc. 2011). These mineralized pods were possibly emplaced along a shear zone (Jooste 1958). The deposit contains 37 Mt of resources and most recent grade estimates show 29.47% $Fe_2O_3$, 10.12% $TiO_2$, 9.25% $P_2O_5$, 0.1% $V_2O_5$, and 0.1% rare earth element oxides (Table 2.1) (IOS Services Géoscientifiques Inc. 2011).

The most recently discovered Fe-Ti-P occurrence examined in this study is Lac Perron, which is located along the western margin of the Lac St. Jean anorthosite suite (Fig. 2.4B). Mineralization at Lac Perron is composed of massive, coarse-grained (cm-sized crystals) nelsonite hosted within coarse-grained anorthosite (Tremblay 2014). Whole rock analysis of several mineralized samples has indicated 64% $Fe_2O_3$, 28.5% $TiO_2$, 6.6% $P_2O_5$, and 0.6% $V_2O_5$ (Table 2.1) throughout the exposed ore body (Tremblay 2014).

The Lac à Paul deposit is the largest deposit within the Lac St. Jean anorthosite suite with estimated resources of 702 Mt at 7.16% $P_2O_5$ (Table 2.1) (Arianne Phosphate Inc. 2013). The ore consists of several zones of high-grade olivine nelsonite and disseminated apatite-, ilmenite-, magnetite-bearing coarse-grained gabbro units which congregate into folds and boudins within the northeastern margin of the Lac St. Jean anorthosite suite (Fig. 2.4B) (Arianne Phosphate 2013).

Along the eastern margin of the study area is the Lac à l'Orignal deposit (Fig. 2.4B), which is hosted within the 1080 ± 2 Ma (van Breeman 2009) Lac Vanel anorthosite rather than the 1.15 Ga Lac St. Jean anorthosite. Glen Eagle Resources Inc. (2012) noted grades of 5.5 % $P_2O_5$ over 45 m within drill core from an apatite-bearing gabbro unit (Table 2.1). It is possible that this deposit is related to the oxide apatite gabbronorite (OAGN) units seen on the eastern margins of the Lac St. Jean anorthosite suite by Hébert et al. (2005).

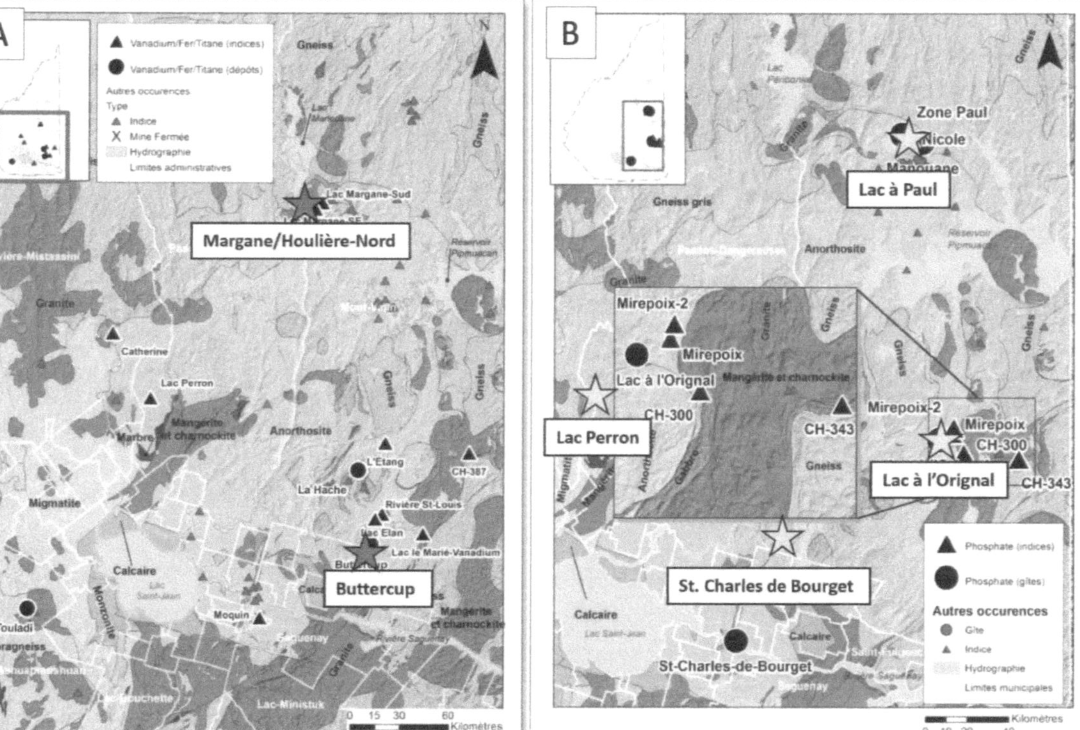


**Figure 2.4:** Locations of all Fe-Ti-V-P mineralization (deposits-gîtes; occurrences-indices) from the Lac St. Jean area: A) Fe-Ti-P mineralization and B) Fe-Ti-V mineralization. Mineralization examined in this study are identified by stars. Modified from CONSOREM, Saguenay Lac St. Jean (2016).

**Table 2.1:** Locations, habit, mineralogy, and grade and tonnage data of the oxide ± apatite mineralization examined in this study. Tonnage data has been obtained from NI-43-101 reports. LSJAS=Lac St. Jean anorthosite suite, Mt=magnetite, ilm=ilmenite, ap=apatite, LVA=Lac Vanel anorthosite, Hm-ilm=Hemo-ilmenite.

| Name | Margin of LSJAS | Morphology | Ore assemblage | Tonnage (Mt) | Fe$_2$O$_3$ (%) | TiO$_2$ (%) | V$_2$O$_5$ | P$_2$O$_5$ (wt%) | References |
|---|---|---|---|---|---|---|---|---|---|
| Buttercup | East | Massive Lense-shaped | Mt>>ilm | 3.2 | 49 | 19 | 0.67 % V$_2$O$_5$ | - | Lalancette (2014) |
| Lac Margane/Houlière | North | Layered sequence within intrusion | Mt ≈ ilm | - | 65 | 25 | 3200 ppm V | - | Tremblay (2016) |
| St. Charles de Bourget | South | Numerous xenolith-bearing lenses | Mt>ilm ± ap | 37 | 48 | 20 | 0.4 % V$_2$O$_5$ | 5.7 | IOS Services Géoscientifiques Inc. (2011) |
| Lac Perron | West | Massive Lense-shaped | nelsonite | - | 64.4 | 25.8 | 0.6 % V$_2$O$_5$ | 6.6 | Tremblay (2014) |
| Lac à Paul (deposit) | North | Layered sequence within intrusion | Ilm>mt ± ap | 702 | - | - | - | 7.16 | Arianne Phosphate Inc. (2013) |
| Lac à l'Orignal | East, Hosted in LVA | Fe-oxide-, ap-rich norite | Hm-Ilm>mt + ap | - | - | - | - | 5.5% /45m | Glen Eagle Resources Inc. (2012) |

# Chapter 3  Description of Fe-Ti-V mineralization

This study is the first to combine outcrop descriptions and detailed petrography of Fe-Ti-V mineralization at Buttercup and Lac Margane/Houlière-Nord, in the Lac St. Jean (LSJ) anorthosite suite. In both locations, the mineralization is composed of massive oxides, and in the case of Lac Margane/Houlière-Nord disseminated oxides as well. Although, the Fe-oxide minerals (magnetite + ilmenite + Al-spinel) are similar, there are significant differences between these two oxide occurrences, described below, related to 1) host rocks, 2) contact relationships with host rocks, 3) grain size of Fe-oxides, 4) modal proportions of Fe-oxide phases, and 5) microtextures of Fe-oxide phases (specifically exsolution textures within magnetite).

Common magnetite exsolution textures observed in oxide mineralization in general (ordered by relative decreasing grain size), are classified as follows (Buddington and Lindsley 1964; Haggerty. 1991; Tan et al. 2017; Arguin et al. 2018); 1) external granules of ilmenite and spinel crystals forming outside host magnetite crystal, 2) internal granules of ilmenite or spinel within magnetite, 3) sandwich exsolutions (coarse-tabular) of ilmenite within magnetite, 4) trellis exsolutions (fine, tabular) of ilmenite aligned along the 111 plane within magnetite, 5) cloth-textured (elongate or granular) crystals of ilmenite and spinel aligned along the 100 plane), and 6) ultra-fine cloth-textured (sub-micron-sized) exsolutions of ilmenite aligned along the 100 plane. Magnetite with exsolution textures of types 3-6 is commonly classified as Ti-magnetite or titanomagnetite (magnetite containing exsolutions of ilmenite $\pm$ ulvöspinel) because of how abundant these exsolutions usually are within the host magnetite. Controls on these exsolution textures remain relatively unconstrained within the Fe-Ti-V-P mineralization of the Lac St. Jean anorthosite suite and thus, will be examined as a part of this study. The petrographic studies presented in Chapters 3 and 4 provide detailed examination of both oxide and silicate textures within the mineralization and the host rocks and should provide insights on mineralogical and textural variation on a regional scale in the Lac St. Jean area. Sample and petrographic information are located in Appendix 1 and summarised below.

## 3.1 Buttercup

The Buttercup Fe-Ti-V deposit, located within the SE corner of the LSJ anorthosite (Fig. 2.4), is composed of three massive oxide lenses (labeled A, B, and C) hosted in anorthosite (Fig. 3.1). The ore bodies are composed of massive coarse-grained Ti-magnetite with trace amounts of ilmenite. Trace amounts of silicate minerals, such as pyroxenes, feldspars, and olivine, have been subject to significant alteration (Lalancette 2014). The first geological work within the area surrounding the Buttercup deposit involved airborne magnetometer surveys over a 3000 square mile area (Goldsmith 1961), which lead to the discovery of massive magnetitite (massive oxide) bodies. The Buttercup deposit is currently exploited for production of dense magnetite aggregates (fairmontresources.ca).

### 3.1.1 Sampling and outcrop descriptions

Samples were taken within approximately a 2 km area surrounding the Buttercup deposit (Figs. 3.1 and 3.2) of which 7 were selected for petrography and geochemical analyses. They comprise massive oxide (n=4), anorthosite at the contact (n=2) and more distal (n=1) of the mineralization. Samples were only taken from the outcropping lenses A and C; none were taken from lense B as it occurs below the surface. One anorthosite (MG-BCP-01a) and one massive oxide (MG-BCP-01b) were sampled along the northern contact of lens A (Fig. 3.3A). Two samples of massive oxide were taken away from the contact in the southern part of lens A (BCP-2 and MG-BCP-03). From lens C, massive oxide was sampled from the basal contact of the lens at the northern edge (MG-BCP-07; Fig. 3.3B) and away from the basal contact (BCP-1). Anorthosite in between the massive oxide lenses A and C (Fig. 3.3E) was sampled (MG-BCP-08). The sample of distal anorthosite (MG-BCP-09) was taken roughly 1 km northwest of the mineralized area (Fig. 3.3F).

Contacts between massive oxide ore and anorthosite are generally sharp (Figs. 3.3A and B). The samples of massive oxide (Fig. 3.3C) are very coarse-grained (>1cm), contain up to 90% magnetite with significantly smaller amounts of coarse-granular ilmenite (5%), spinel (5%), and highly altered Fe-Mg silicates (<5%). There is no visible difference in textures in hand specimen in samples taken from near the contact and within the central part of the ore lense. Anorthosite near the contacts of the ore body are locally megacrystic (Fig. 3.3D). These anorthosites are composed of close to 100 modal% plagioclase with trace amounts of Fe-oxides which show a weakly- to moderately-developed foliation, and locally garnet crystals (Fig. 3.3E). Distal anorthosite is very coarse-grained with only a local presence of Fe-oxide and orthopyroxene crystals. It is locally megacrystic, showing a moderately developed foliation and granoblastic texture.

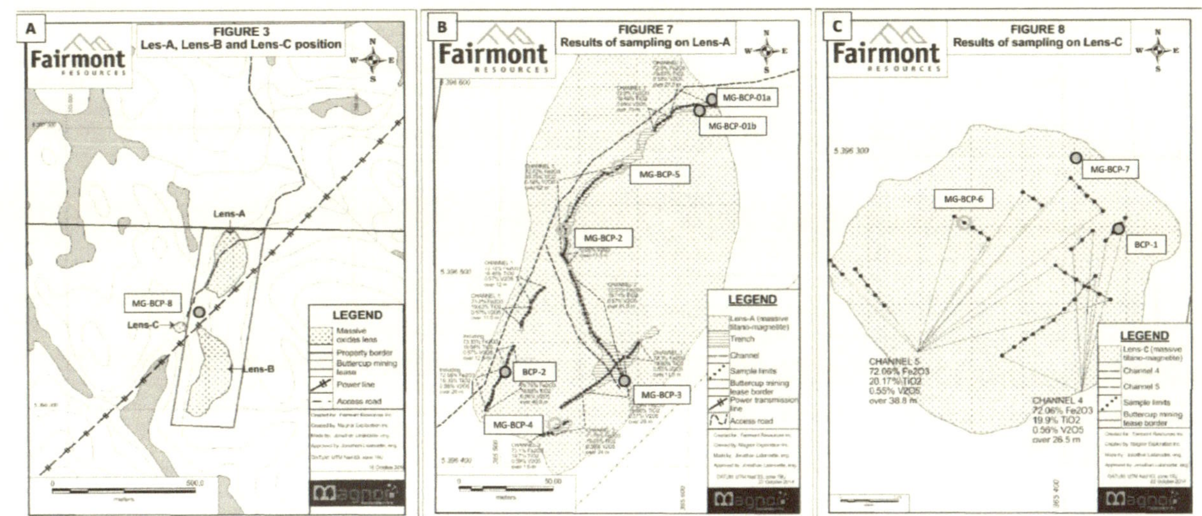


**Figure 3.1**: Maps taken from Lalancette (2014), by Fairmont Resources inc, displaying the exposed outline and locations of massive Fe-oxides bodies of the Buttercup Fe-Ti-V deposit (A) and channel sampling locations and Fe-Ti-V grades from lense A and lense C (C), which outcrop on surface. Lense B is defined by drilling and geophysics. Locations of samples (collected = open circles; analysed = filled circles) are overlain on the outline of the massive Fe-oxide lenses

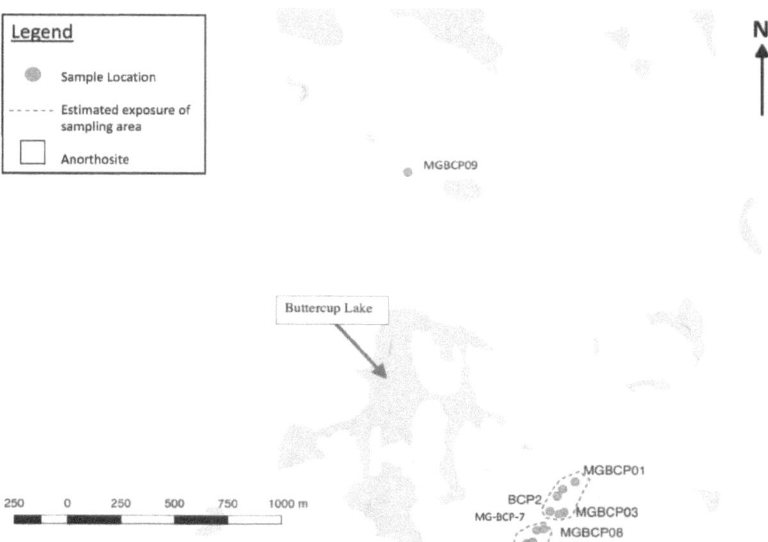


**Figure 3.2.** Map displaying locations of all samples from the Buttercup deposit and surrounding area collected in this project. Note MG-BCP-09 was taken from an anorthosite outcrop located roughly 1km away from the deposit. Base layer of the map from ©2018 Google.

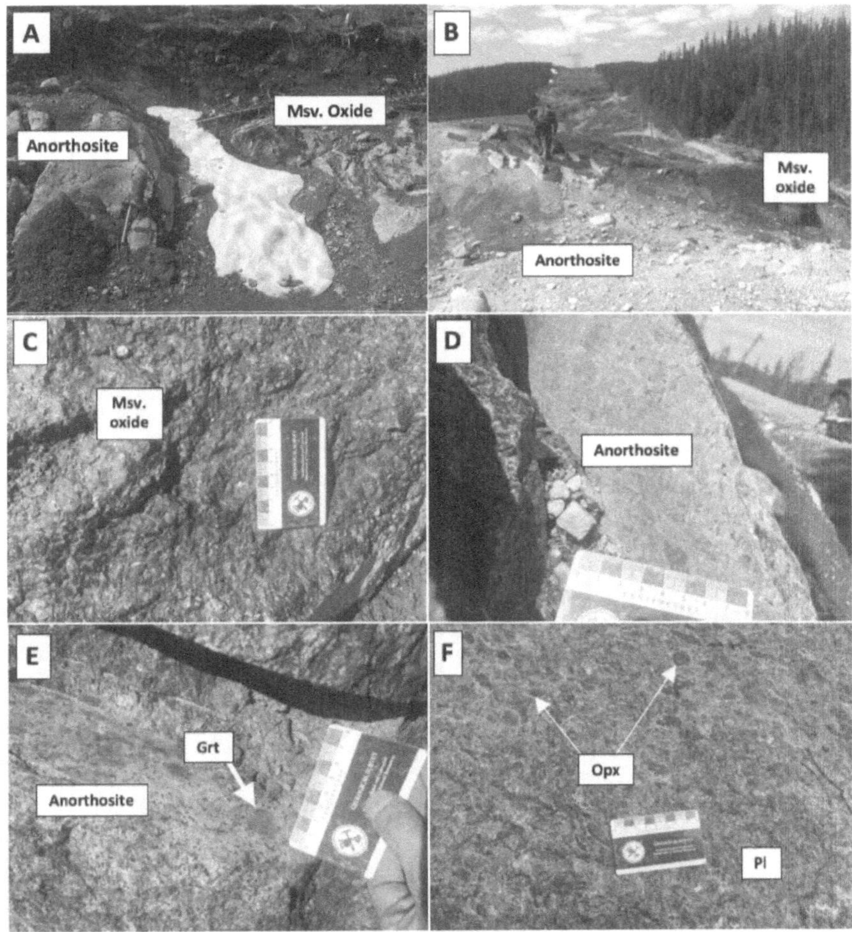


**Figure 3.3.** Outcrop photographs displaying field relationships of the Buttercup Fe-Ti-V deposit and surrounding host anorthosite. A) Contact between anorthosite (MG-BCP-01a) and massive (Msv.) oxide (MG-BCP-01b) at the northern contact of lens A, hammer for scale. B) Contact between anorthosite and massive oxide at the northern contact of lens C, geologist for scale. C) Massive oxide ore from lens A. D) Anorthosite near the northern contact of lens A. E) Anorthosite in between lenses A and C (MG-BCP-08) containing minor garnet (Grt). F) Anorthosite/leuconorite (Pl-plagioclase; Opx-orthopyroxene) located roughly 1km north-west of mineralization (MG-BCP-09).

3.1.2    Petrography

The 2 main lithologies from the Buttercup Fe-Ti-V deposit are massive oxide and anorthosite. Both show significant textural differences microscopically with respect to distance from the mineralization/anorthosite contact.

The modal mineralogy of massive oxide (Fig. 3.4A) mineralization remains relatively constant at the deposit-scale. Magnetite is the dominant opaque mineral, ranging from 80-90%, with minor ilmenite and Al-spinel both present at around 5-10% modality each. Very fine-grained sulfides, mainly chalcopyrite and pyrrhotite, occur in trace amounts. Magnetite crystals in these samples are very coarse-grained (>1cm) and display a variety of exsolution textures (Fig. 3.4). Ilmenite is present as both as internal (cloth or sandwich) and external exsolution (coarse-grained granule) phases within magnetite. Al-spinel is present mostly as coarse granules (>1cm), as cloth-textured exsolutions within magnetite, or as inclusions within tabular ilmenite crystals. No primary silicates were preserved. Although the mineralogy is similar for all massive oxides at Buttercup, microtextures vary considerably between those sampled away from the contacts and those at the contact with the host anorthosite.

Within samples taken away from the contact, exsolution phases in magnetite are predominantly 1) ultra-fine-grained cloth-textured ilmenite/ulvospinel (<5 μm), with 2) minor fine-grained, subhedral internal granules of Al-spinel (<20 μm), and 3) discontinuous, linear, ultra-thin, elongate (100 μm long but, their width is <5 μm) trellis-style exsolutions of Al-spinel (Fig. 3.4B). Both the granular and linear exsolutions of spinel are separated from the ilmenite exsolutions by 5 μm wide exsolution-poor zone (Fig. 3.4B). These textures can only be seen at high resolution. Sandwich exsolutions of ilmenite within magnetite and external granule exsolutions of ilmenite are uncommon in massive oxides away from the contact. External granule exsolutions of Al-spinel (mm to cm-sized) are present but are sometimes difficult to observe even at low resolution as magnetite very coarse-grained (Figs 3.4A, C, and D).

Exsolution phases within the massive oxides near the contact of host anorthosite (Figs. 3.4C, D, and E) are slightly more abundant and much coarser-grained than exsolution phases in massive oxides away from the contact. The dominant exsolutions styles within massive oxides near the contacts are 1) external granule exsolutions of both ilmenite (more common in lense A) and Al-spinel (more common in lense C) and 2) sandwich exsolutions of ilmenite, 50-100 μm wide and >500 μm long (Fig. 3.3D). The external granule exsolutions of ilmenite and Al-spinel are anhedral possibly because crystal growth follows the magnetite grain boundaries. In addition, trellis-style exsolutions of both ilmenite and spinel are >5 μm wide and 50-100 μm long and intersect at roughly 60° and 120° angles to each other (Fig. 3E). The same magnetite crystal can host both trellis ilmenite and spinel but the angle of intersection for the spinel exsolutions are organized in opposite directions to that of ilmenite. Trellis-style exsolutions appear to be absent from massive oxide samples located near the contacts at lense C; the sole internal exsolution phase is fine-grained spinel granules (Fig. 3.4F).

Silicates present as trace amounts within the massive oxides display highly complex secondary textures. They consist of of symplectic intergrowths of nepheline and olivine (determined by EMPA: Chapter 5) surrounded by a corona of amphibole ± biotite (Fig. 3.5A). Al-spinel is present within the corona as inclusions within the amphibole. Symplectic cores of these silicate patches are rarely preserved in most cases. In most cases, only the amphibole crystals are present.

All anorthosites sampled near the contacts with the oxide bodies are relatively similar in mineralogy, containing 90-95% plagioclase and approximately 5-10% Fe-oxides (both magnetite and ilmenite) as primary mineral phases. Disseminated magnetite and ilmenite appear to be present in relatively equal proportions and are much finer-grained than Fe-oxides within massive oxide and are partially altered in the presence of silicates (Fig. 3.5B). Magnetite within the host anorthosites has significantly fewer exsolutions than magnetite within massive oxide (Fig 3.5B). Plagioclase crystals are predominantly coarse-grained, anhedral prisms displaying local deformation in albite twins through pinch and swell textures and bending (Fig. 3.5C). These coarse crystals can locally be surrounded by fine-grained plagioclase crystals which display moderately- to well-developed granoblastic texture, especially within proximity to Fe-oxides (Fig. 3.4C). This is described as protoclastic texture, a texture noted in Proterozoic anorthosites by Ashwal (1993) which is typically caused by high temperature deformation before complete crystallization. However, in some cases protoclastic texture could also be a product of solid state deformation during localised deformation and regional metamorphism.

Possible secondary mineral phases within these contact anorthosites include biotite, amphibole, and garnet. The Fe-oxides are surrounded by a simple biotite corona (MG-BCP-01a) or by a wide inner biotite corona that can be partially or completely surrounded by an outer corona of amphibole (MG-BCP-08: Fig. 3.4D). Garnet is present and forms coarse, anhedral masses within this sample, containing abundant inclusions of plagioclase, Fe-oxides, and biotite (Fig. 3.4E).

Distal anorthosite (MG-BCP-09) appears to be coarser-grained and less deformed, i.e., granoblastic texture less well developed, than anorthosite near the contacts with mineralization (Fig. 3.3F). Fe-oxides are also present in trace amounts and show consumption by complex metamorphic/re-equilibration reactions, such as magnetite + ilmenite symplectites

surrounded by orthopyroxene, hornblende, and biotite (Fig. 3.5F).

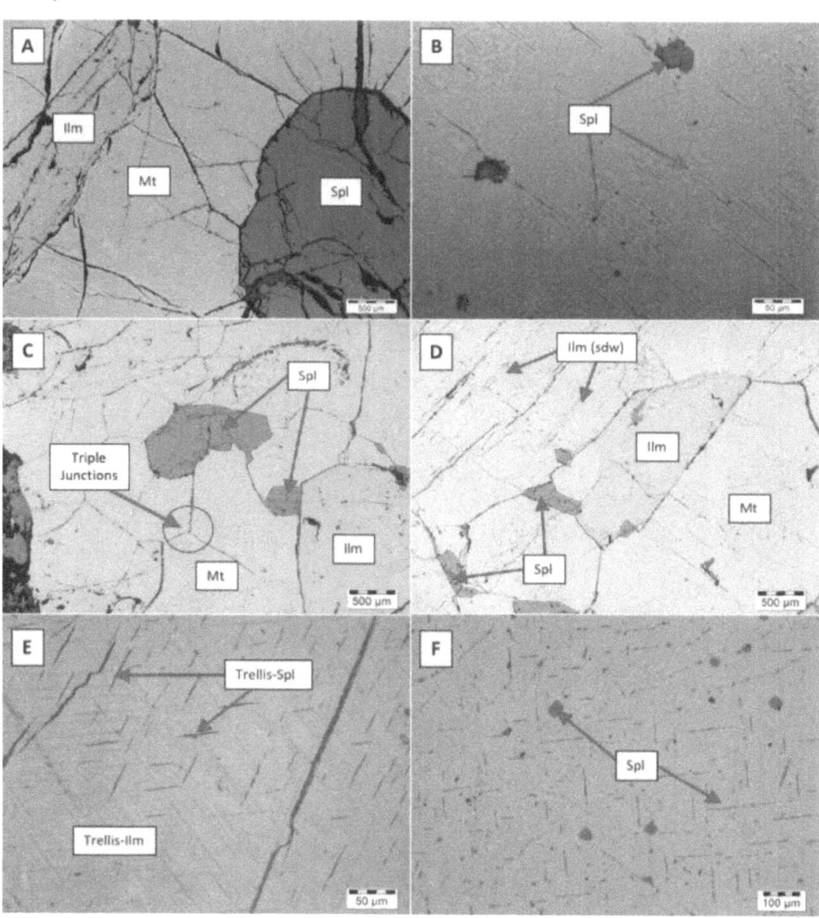



**Figure 3.4** Photomicrographs displaying microtextures of Fe-oxides within the Buttercup deposit, note the overall coarsening of exsolutions from centre of ore body (A-B) towards the contact with host anorthosite (C-F). A) Low resolution image of magnetite (mt), ilmenite (ilm) and Al-spinel (spl) within massive oxides from center of the ore body (MG-BCP-3). B) Ultra-fine cloth-textured ilmenite and granular Al-spinel (spl) exsolutions within magnetite (MG-BCP-7). C) Dynamic recrystallization and grain size reduction of magnetite (MG-BCP-07), also note external granule exsolution of spinel along the grain boundaries of magnetite which can locally generate coarse-grained Al- spinel. D) Magnetite displaying external granule exsolutions of Al-spinel and ilmenite (MG-BCP-7) along with sandwich- (sdw) textured ilmenite exsolutions, note the decreasing size and quantity of exsolutions towards grain boundaries, this is absent when external granule exsolution does not occur. E) Trellis textured exsolutions of ilmenite and spinel within magnetite (MG-BCP-01b). F) Cloth-textured ilmenite exsolutions occurring with stringer and granular spinel exsolutions within magnetite (BCP-2).

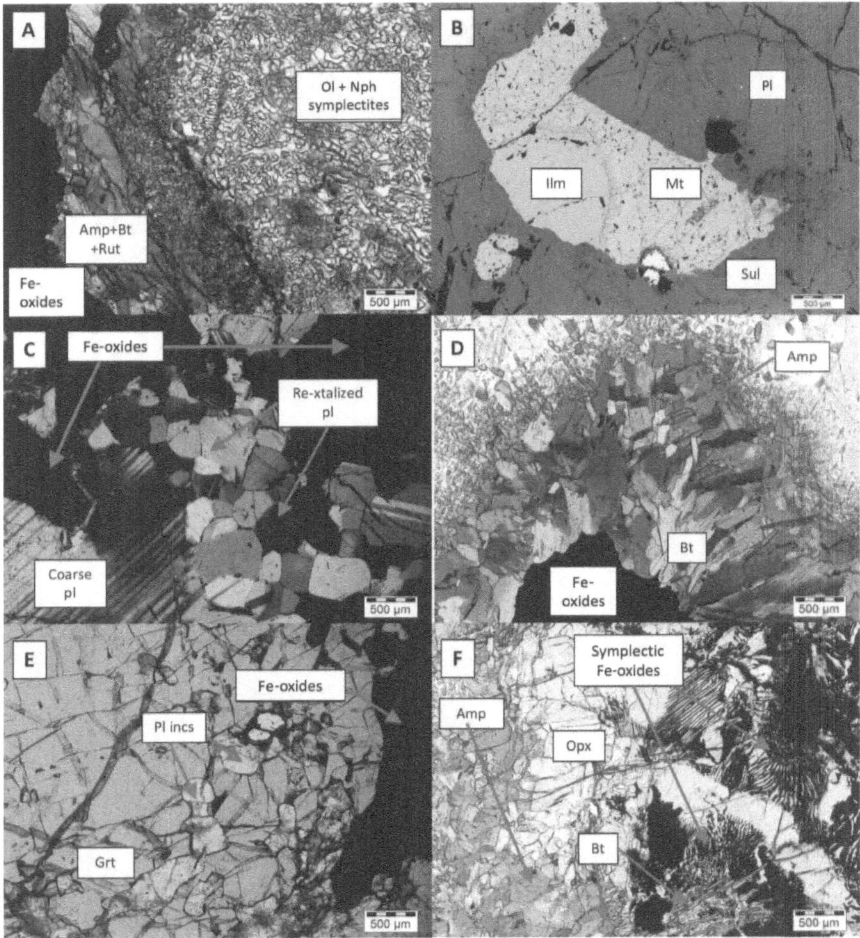



**Figure 3.5.** Photomicrographs of silicate minerals within massive ore (A) and host anorthosite (B-F). A) Silicate minerals associated with the massive ore (BCP-2); within the core there are symplectites between olivine (Ol) and nepheline (nph) surrounded by a corona of amphibole (amp), biotite (bt), and rutile (rut) in contact with Fe-oxides. B) Disseminated magnetite and ilmenite occurring with a fine-grained sulfide (sul) mineral located within contact anorthosite (MG-BCP-01a). C) Protoclastic texture within anorthosite shown by fine-grained granoblastic-textured plagioclase (pl) in contact with coarse plagioclase crystals and Fe-oxides from contact (MG-BCP-01a). D) Double-layered corona around Fe-oxides with an inner corona of biotite and a moderately developed outer corona of amphibole (MG-BCP-08). E) Anhedral, coarse-grained garnet (grt) containing inclusions (inc) of plagioclase and Fe-oxides (MG-BCP-08). F) Symplectic Fe-oxides surrounded by orthopyroxene (opx), amphibole and biotite (MG-BCP-09).

## 3.2   Lac Margane/Houlière-Nord

The Lac Margane and Lac Houlière-Nord Fe-Ti-V occurrences are located within the northern margin of the Lac St. Jean anorthosite suite, 30-40 km west of the Lac à Paul Fe-Ti-P deposit. Mineralization at these occurrences consists of multiple discontinuous, decameter- to meter-scale bands of massive Fe-oxides interlayered with a norite unit containing disseminated Fe-oxides (Tremblay 2016; this study). All Fe-oxide occurrences within the Lac Margane/Houlière-Nord area are assumed to be genetically related as they generally occur along strike of each other and are relatively parallel to the anorthosite/norite boundary defined by Hébert et al. (2009) (Fig. 3.6).

The first company that worked on the area was NQN mines in 1970 who carried out geophysical surveys, regional mapping, and drilling programs (see review in Tremblay 2016). Further geophysics, trenching, drilling, and regional mapping were performed by Soquem and Virginia (Roy 2001), and then again by Arianne resources Inc. (Tremblay and Lefebvre 2013). The most detailed regional mapping of the area was completed by the Québec Ministry of Energy and Natural Resources (Hébert et al. 2000; 2009; 2010) who also viewed and described the deposits. The most recent prospecting work to date is geochemical sampling from surface outcrops performed by Tremblay (2016), whose data is given in Table 3.1 for each of the occurrences within the Margane/Houlière area.

3.2.1    Sampling and outcrop descriptions

A total of 20 samples used for this project come from four occurrences within the Margane/Houlière area; 1) Lac Margane-Est (n=13), 2) Lac Margane-Sud-Est (SE) (n=4), 3) Lac Margane-Sud (n=1), and 4) Lac Houlière-Nord (n=3). All samples taken by Tremblay (2016) contain Fe-oxides that were taken at certain points along traverses perpendicular to the Fe-oxide layers to collect adequate samples of both massive Fe-oxides and Fe-oxide-rich norites. Locations of these samples are reported in (Fig. 3.6; Table 3.2) and were selected for further analysis in this study. In addition, host rocks of Lac Margane-Est were taken during my field visit in 2017. These were originally classified as norites by (Hébert 2000), but all orthopyroxene has been converted to amphibole and should contain the prefix (meta)norite. Locations of these samples are indicated in Fig. 3.7.

The exposed mineralization at the Lac Margane-Est area consists of flat lying, poorly exposed outcrops (Fig 3.9A) consisting of decameter- to meter-scale layers of massive oxide within the host norite which is crosscut by approximately 15 cm wide syenitic dikes (Fig 3.9B). Contacts between the massive oxide layers and norites appear to be gradational as the Fe-oxides within the norite show a gradual decrease proportional to distance away from the massive oxide layer (Fig. 3.9B). Both the massive oxide layer and the host norite display a well-developed foliation. In the massive oxides, it is defined by thin (approximately 3mm wide) discontinuous layers of amphibole and biotite (Fig. 3.9C). Within the norites, garnet can locally form coronas around Fe-oxides. Host rocks taken from within the exposed mineralization at Lac Margane-Est represent anorthositic (MG-LMG-01) and leuconorite (MG-LMG-02) layers. Both samples contain disseminated Fe-oxides.

Approximately 250 m north of the exposed mineralization, the host norite is equigranular and coarse-grained (MG-LMG-03a) but can locally be megacrystic (>10 cm) comprising plagioclase and hornblende (MG-LMG-03b: Fig. 3.9D). The foliation in this

norite is poorly developed relative to that of the norite in contact with the massive oxides (Fig.

3.9E). Norite approximately 2 km north of the exposed mineralization is equigranular,

relatively unaltered and unmetamorphosed and contains relict orthopyroxene (MG-LMG-04:

Fig.3.9F).

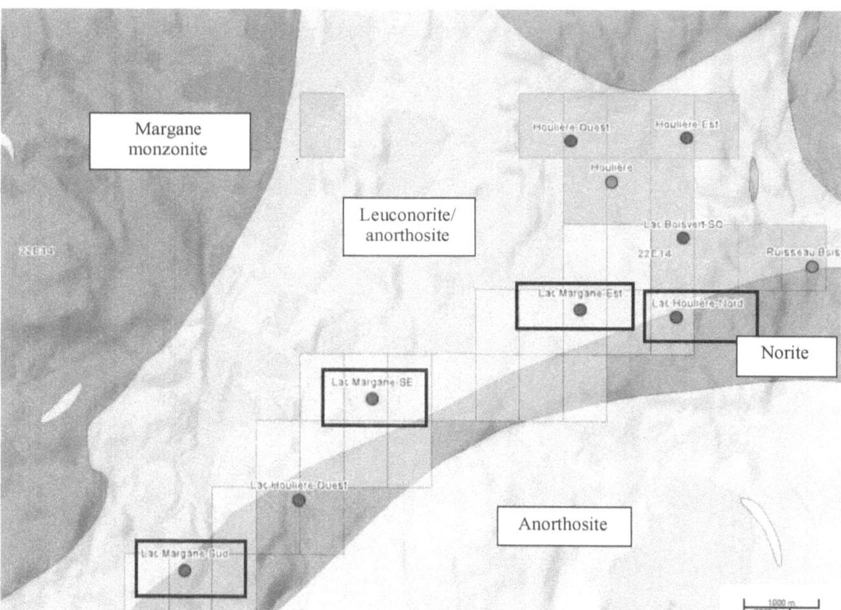


**Figure 3.6:** Geological map (map sheet 22E14) displaying the locations of massive Fe-oxide showings within the Margane/Houlière area and the surrounding rock units (Tremblay 2016). Note that most deposits occur near the contact between the anorthosite/leuconorite and gabbronorite units. Black rectangles indicate sampling areas Bedrock mapping by Hébert (2000). Blue dots=Fe-oxide mineralization, Brown dots=Ni mineralization, Orange dots=Cu mineralization.

**Table 3.1:** Average whole rock geochemical values of $Fe_2O_3$, $TiO_2$, $V_2O_5$, and $Cr_2O_3$ for massive oxide layers from each occurrence studied during this project. Data from Tremblay (2016).

|  | $Fe_2O_3$ % | $TiO_2$ % | $V_2O_5$ % | $Cr_2O_3$ % |
|---|---|---|---|---|
| Lac Margane-Sud | 66.94 | 16.98 | 0.559 | 1.63 |
| Lac Margane-SE | 54.70 | 21.84 | 0.407 | - |
| Lac Margane-Est | 57.66 | 14.93 | 0.512 | 1.79 |
| Lac Houlière-Nord | 62.60 | 23.13 | 0.544 | - |

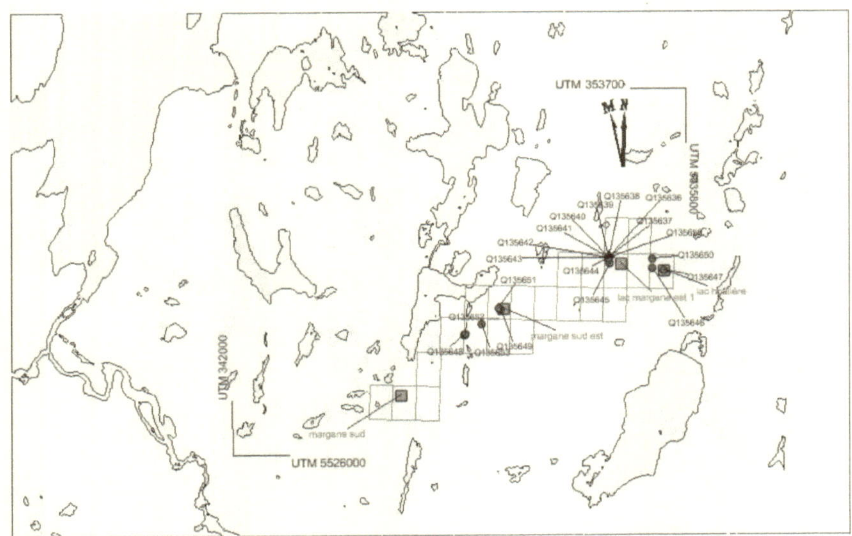


**Figure 3.7:** Map showing the locations of samples taken by Tremblay (2016) during prospecting work. Corresponding sample names are located in Table 3.2.

**Table 3.2:** Table showing the sample numbers and locations collected by Tremblay (2016) which were studied in further detail in this study. This Table corresponds with Fig. 3.7. Locations are given using projection NAD83/UTM zone 19N. Each sample is described in Appendix A. Information for host rock samples collected during fieldwork for this project is located in Appendix A.

| Sample # | UTM Easting | UTM Northing | Lab code (Tremblay 2016) |
|---|---|---|---|
| MA-01a | 351732 | 5531293 | Q135636 |
| MA-01c | 351732 | 5531293 | Q135638 |
| MA-01e | 351732 | 5531293 | Q135640 |
| MA-01g | 351732 | 5531293 | Q135642 |
| MA-02a | 351766 | 5531166 | Q135645 |
| Houlière Nord | 353176 | 5530982 | Q135647 |
| MA08 | 348001 | 5529241 | Q135648 |
| MA3b | 348950 | 5529917 | Q135649 |
| WP207 | 352887 | 5531263 | Q135650 |
| MA3a | 348893 | 5529937 | Q135651 |
| WP246 | 348013 | 5529260 | Q135652 |
| MA07 | 348443 | 5529522 | Q135653 |

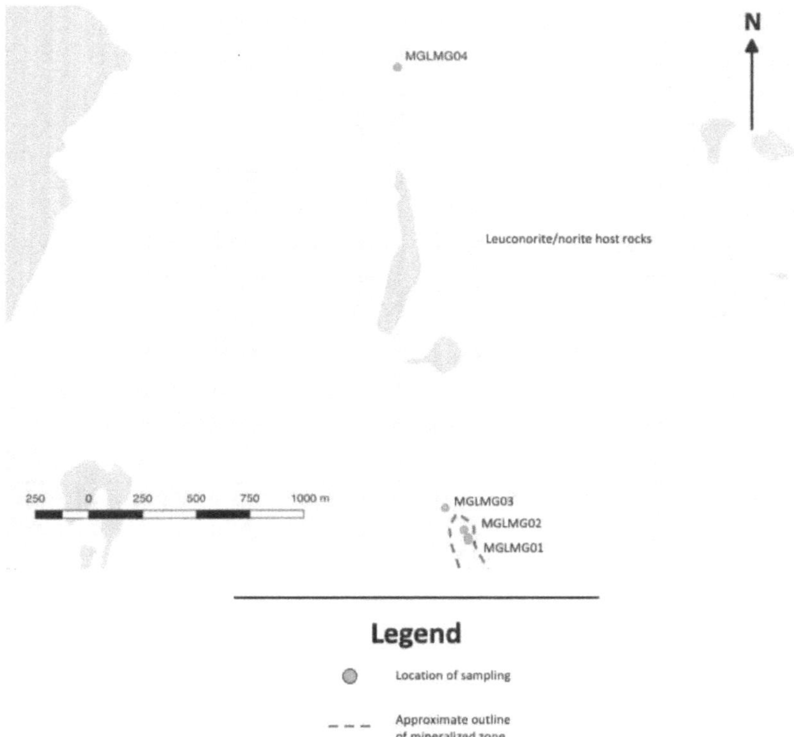


**Figure 3.8:** Map showing the locations of the host rock samples taken for this project from Lac Margane-Est during the 2017 field season. Base layer of the map from ©2018 Google.

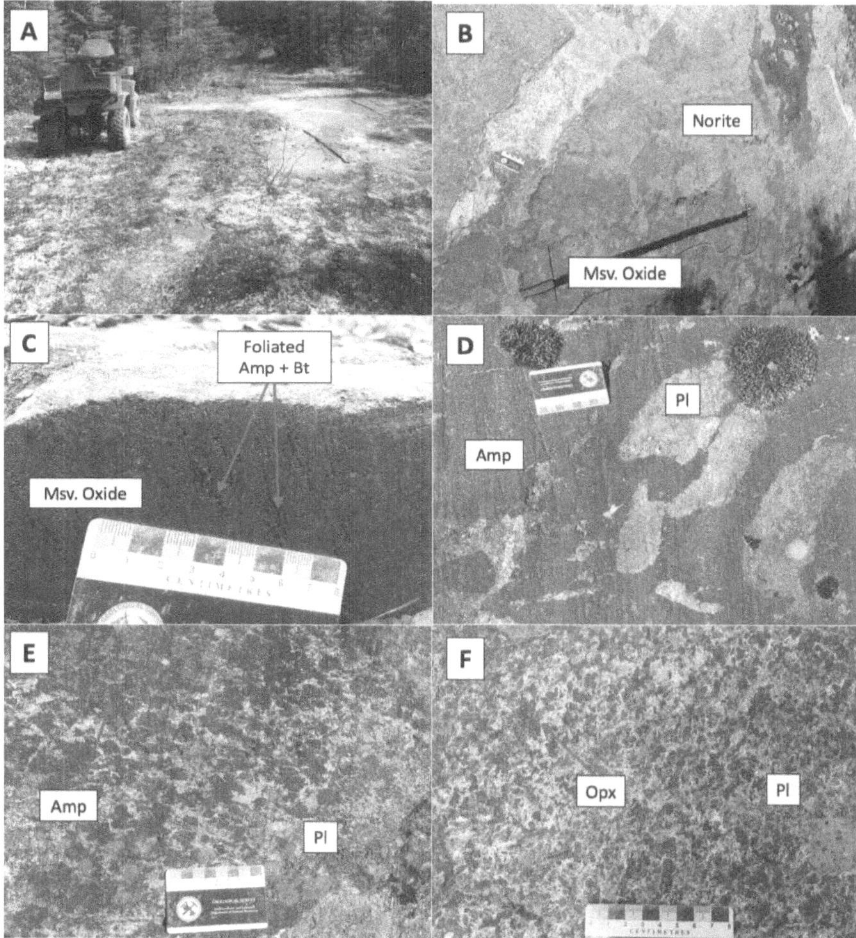


**Figure 3.9:** Outcrop photographs of the exposed mineralization (A-C) and host noritic rocks (D-F) from the Lac Margane-Est area. A) Moss-strip exposure of the mineralized zone, ATV for scale. B) Massive (Msv.) oxide layer showing a slight gradational contact with the host norite, both the norite unit appears to be crosscut by a felsic dyke. C) Close-up photograph of the massive oxide, note the well-developed foliation and thin bands of amphibole (amp) and biotite (bt). D) Amphibolitized norite displaying megacrystic plagioclase (pl) and amphibole (MG-LMG-03b). E) Equigranular amphibole norite from the same outcrop as D (MG-LMG-03a). F) Equigranular norite with relict orthopyroxene crystals (MG-LMG-04).

37

3.2.2   Petrography

Between the Lac Margane and Lac Houlière-Nord occurrences, a total of 18 polished thin sections were described from the mineralised zones. This includes Lac Margane-Est (n=9), Lac Margane-Sud-est (n=4), Lac Margane-Sud (n=2), and Lac Houlière-Nord (n=3). The predominant lithologies are massive oxide ore (n=4) and Fe-oxide-rich (meta-)norite (n=13). These lithologies are similar within each occurrence but can display minor textural differences. Massive oxides generally contain >90% Fe-oxides, where magnetite (40-65 modal%) is slightly dominant over granular ilmenite (45-30 modal%). Coarse-grained Al-spinel is locally enriched in these samples as modal percentages can range from 5-20%. Trace silicates (<5%) present within the massive ore include amphibole and biotite. Fe-oxide-rich norites typically contain 5-40 modal% Fe-oxides (semi-massive to disseminated oxides), where ilmenite (5-20 modal%) predominates over magnetite (0-10%); spinel can be present either in trace amounts (<5%) or up to 15 modal%. Assuming hornblende is a secondary phase due to the alteration of either ortho- or clinopyroxene during amphibolite-facies metamorphism, the only primary igneous mineral within this lithology is plagioclase. However, garnet can be locally present in significant amounts (5-20%) within some samples and trace amounts of biotite are present with most of these samples. Despite the significant mineralogical differences between several of these samples, they are meta-norites because 1) we assume that this was the primary lithology before metamorphism/re-equilibration, 2) no relict pyroxenes were observed within the polished thin sections thus hindering the ability to successfully distinguish between gabbro or norite, and 3) previous regional mapping in the area by Hébert (2000) described the rocks surrounding the Margane/Houlière deposits as norites.

The distribution of the Fe-oxide minerals in massive ores is non-uniform as magnetite and ilmenite are present as grouped clusters of approximately 1 mm-sized grains which displaying moderately- to well-developed dynamic recrystallization shown by triple junctions within a single polished thin section (Fig. 3.10A). Al-spinel is present in these samples as singular, coarse, sub- to anhedral external granules or as an internal exsolution phase. While ilmenite does not contain any exsolution phases, magnetite displays 3 different styles of exsolution; 1) external granule exsolutions (approximately >200 μm) of ilmenite and/or Al-spinel, 2) granular exsolutions of <20 μm Al-spinel arranged in a trellis pattern, and 3) >100 μm sandwich ilmenite (rare within these occurrences). External granule exsolutions are the most complex style as they can be classified further as 1) granular spinel, 2) tabular ilmenite located between two magnetite crystals (Fig. 3.10B), and 3) triagonal-shaped ilmenite crystals located within triple junctions of magnetite crystals (Fig. 3.10C). Exsolution styles 2 and 3 can also occur internally within magnetite crystals. Granular spinel exsolutions (style 2) can be locally coarse (~100 μm) but, are generally much smaller (<20) and arranged in a linear pattern along the {111} plane in magnetite. Lastly, sandwich exsolutions of ilmenite within magnetite can be up to 500 μm long and 200 μm wide and have serrated crystal faces (Fig. 3.10D). Locally, fine-grained (<10 μm) Al-spinel is present at the boundary of the exsolved ilmenite and the magnetite host. Exsolutions are also present within coarse, anhedral spinel crystals (Fig. 3.8E); they are indistinguishable due to their very-fine grain size, which is too small to be examined by EMPA (Fig. 3.8F). Exsolutions within both magnetite and spinel are concentrated within the core of the crystal as they tend to disappear towards the edge of the crystal. This is possibly an effect of external granule exsolution.

Relative to massive oxides, Fe-oxide mineral habits differ within the disseminated Fe-oxide-rich norites as 1) Fe-oxides are finer-grained, 2) ilmenite becomes more abundant than magnetite, 3) spinel shows a dramatic decrease in abundance, and 4) magnetite lacks exsolution phases. When samples contain approximately 20 - 40% oxides (heavily disseminated to semi-massive), deformation is predominantly displayed by dynamic recrystallization (Fig. 3.11A). When there are <20% Fe-oxides (disseminated) within a sample, Fe-oxides display a well-developed foliation and elongation of singular crystals (Fig. 3.11B).

Deformation and re-equilibration within the Fe-oxide-rich norites are best illustrated by the silicate phases. Dynamic recrystallization is most developed within plagioclase relative to amphibole as plagioclase crystals are much finer-grained and display more mature triple junctions than amphibole crystals (Fig 3.11A). Amphibole displays two main crystal habits, each with a dark green to yellowish green pleochroic scheme. They are present as either 1) coarse-grained amphibole with moderately-developed triple junctions within samples containing 40% Fe-oxides (Fig. 3.11C) and 2) tabular and elongate amphibole defining a foliation within samples containing <20% Fe-oxide (Fig. 3.11D). Both crystal forms can locally contain inclusions of both Fe-oxides and plagioclase further implying that it is a secondary mineral phase. Tabular biotite crystals, that display a deep reddish-brown to pale brown pleochroic scheme, occur locally either within or near amphiboles and Fe-oxides (Fig. 3.11D). Garnet is locally present within Fe-oxide-rich norites (e.g., samples LHN-1, Houlière-Nord, MA-01g, and MA-01c) and occurs as anhedral masses which can locally be coarse-grained enough to cover an entire thin section (cm-scale). These anhedral masses contain abundant inclusions of amphibole and sometimes biotite indicating possible formation due to metamorphic/re-equilibration reactions between Fe-oxides and plagioclase ± amphibole (Fig. 3.11E). Apatite is only present in trace amounts (<1%) within sample MA-01a as subhedral grains roughly 500 µm in diameter (Fig. 3.11F), however, is absent in all other samples. Apatite has never been reported in any concentration within the Margane/Houlière mineralization.

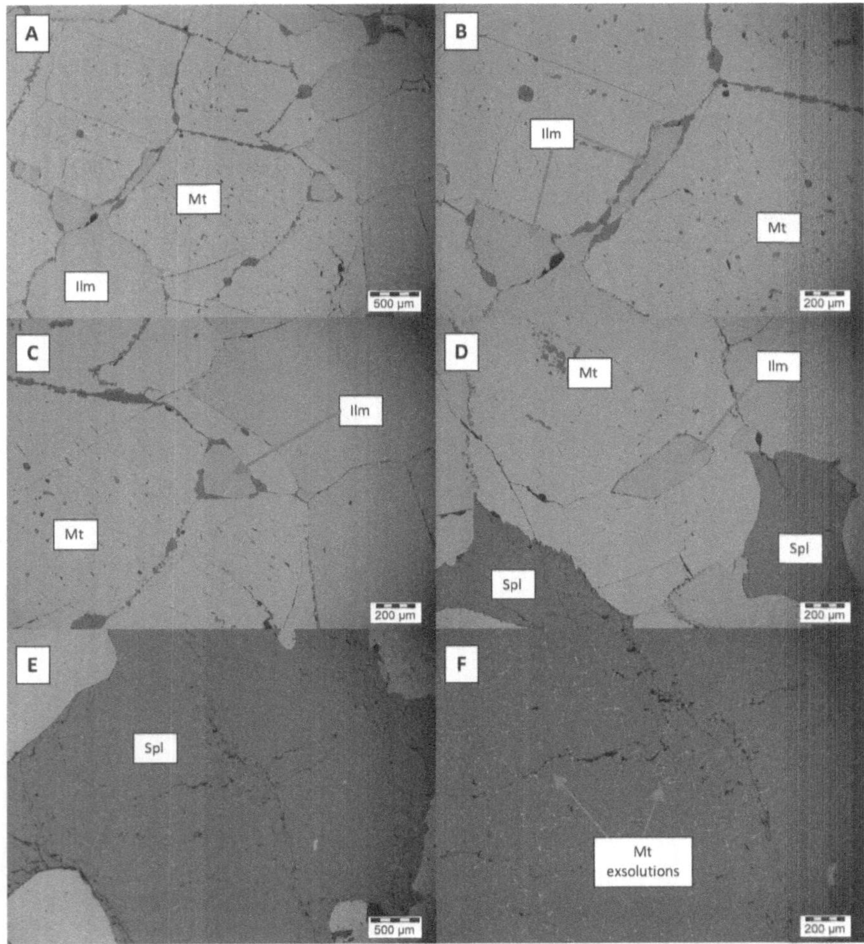

**Figure 3.10:** Photomicrographs displaying microtextures of Fe-oxides within massive ores from the Lac Margane/Houlière deposits. All photomicrographs taken from sample LMG-1. A) Typical massive oxide ore with magnetite (mt) showing both internal and external granule exsolution of spinel (spl), local external granule exsolution of ilmenite (ilm), and fine-grained trellis spinel. B) Same area as Fig. 3a but at higher resolution to emphasize external granule exsolution of ilmenite along a parallel grain boundary between 2 magnetite crystals. C) Same area as Fig. 3a but at higher resolution to emphasize external granule exsolution of ilmenite within a triple junction between magnetite crystal, the ilmenite grain contains a partial rim of spinel. D) Internal exsolutions of ilmenite within magnetite. E) Coarse, anhedral spinel crystal displaying fine-grained exsolutions of magnetite, note these exsolutions are concentrated within the core of the crystal. F) High resolution view of magnetite exsolutions within Al-spinel.

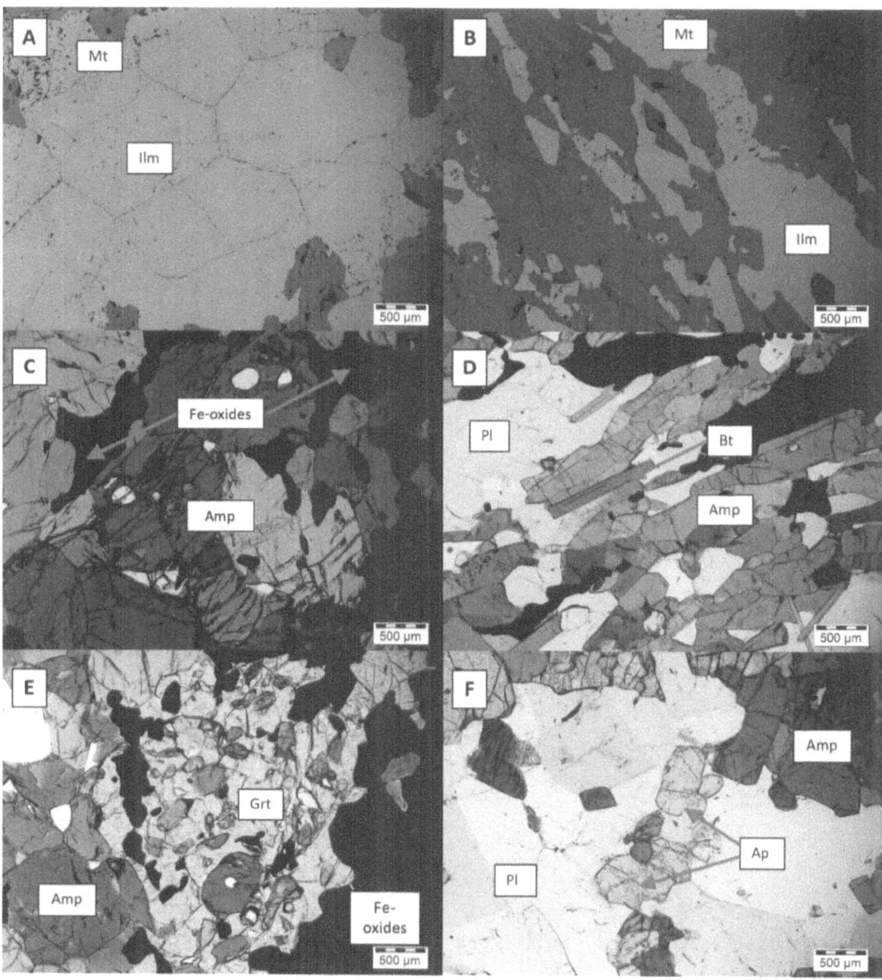


**Figure 3.11:** Photomicrographs of Fe-oxide and silicate textures within samples containing semi-massive and disseminated Fe-oxide-rich norite host, note the predominance of ilmenite (ilm) over magnetite (mt) in A and B. A) Ilmenite and magnetite displaying triple junctions within semi-massive oxide (MA-01a). B) Foliated and elongated ilmenite + magnetite within disseminated oxide (MA-02a-a). C) Coarse-grained amphibole (amp) displaying triple junctions and inclusions of plagioclase (pl) within semi-massive oxide (MA-03b). D) Strongly foliated amphibole including Fe-oxides and minor biotite (bt) and plagioclase within disseminated oxide (MA-02a-a). E) Fe-oxide norite containing garnet (grt) with multiple inclusions of Fe-oxides and amphibole (MA-01g). F) Trace apatite (ap) found locally within Fe-oxide norite (MA-01a).

# Chapter 4 Descriptions of Fe-Ti-P mineralization

Outcrop descriptions and detailed petrography of three occurrences of Fe-Ti-P mineralization (St. Charles de Bourget, Lac Perron, and Lac à Paul) within the 1170-1140 Ma Lac St. Jean anorthosite suite and one occurrence (Lac à l'Orignal) within the 1080 Ma Lac Vanel anorthosite are described in this Chapter. The Fe-Ti-P mineralization either occurs as 1) nelsonites hosted by anorthosites (i.e. St. Charles de Bourget and Lac Perron) or 2) semi-massive to disseminated Fe-oxides + apatite within mafic/ultramafic lithologies (i.e. Lac à Paul and Lac à l'Orignal). Mineralization at St. Charles de Bourget and Lac à Paul occurs within proximity to ferrodiorite dykes while no such lithologies have been reported in association with Lac Perron or Lac à l'Orignal. Similar to the Fe-Ti-V mineralization, the main differences among the Fe-Ti-P mineralization are the; 1) host rocks, 2) contact relationships with host rocks, 3) grain size of Fe-oxides and apatite, 4) modal proportions of Fe-oxide phases and apatite, and 5) microtextures of Fe-oxides. Although microtextures of Fe-oxides vary among the Fe-Ti-P mineralization, the textures remain relatively similar to those noted within the Fe-Ti-V samples. However, mineralization at Lac à l'Orignal contains hemo-ilmenite, which is characteristic of younger anorthosites and their mineralization, making it different to mineralization within the Lac St. Jean anorthosite suite. Overall, detailed descriptions of Fe-oxide textures within all lithologies of Fe-Ti-P deposits of the Lac St. Jean area should provide insights on mineralogical and textural variations on a regional scale. Sample and petrographic descriptions are located in Appendix 1.

## 4.1   St. Charles de Bourget

Mineralization at St. Charles de Bourget is located within the southern part of the Lac St. Jean anorthosite suite near the Saguenay River and is located 2-3 km SE of the Lac Chabot ferrodiorite (Côté 1986, Fig. 4.1). The first notion of Ti-magnetite bodies within the vicinity of the St. Charles de Bourget township was by Laflamme (1882). The southern portion of the mineralization comprises multiple large pods of massive Ti-magnetite within anorthosite (Denis et al. 1913; Dulieux 1915). The most relevant mapping of mineralization within the St. Charles de Bourget area available was performed by Osbourne (1944); Jooste (1948, 1958); and Bachari (2004) who noted that Ti-magnetite (Fe-Ti±V) bodies were concentrated in the south while nelsonite bodies (Fe-Ti-P) were concentrated in the north of an approximate area of a few 100 m$^2$ (Fig. 4.1). These oxide bodies were thought to have been emplaced along a brittle-ductile shear zone based on the presence of mylonites along the ore-anorthosite contact (Jooste 1958).

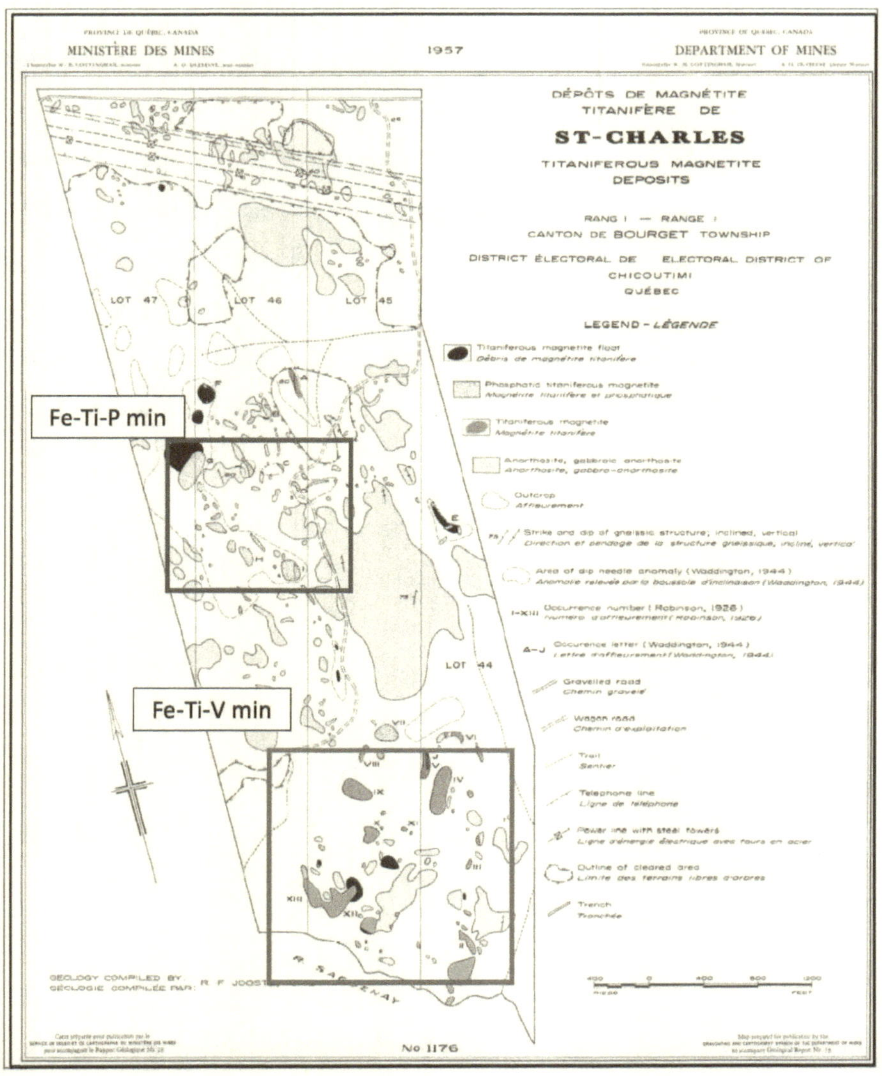


**Figure 4.1:** Geological map of the St. Charles de Bourget area (Jooste et al. 1958). Note that anorthosite-hosted Fe-Ti±V mineralization predominantly in the south (black box) and Fe-Ti-P mineralization in the north (red box).

46

4.1.1    Sampling and outcrop descriptions

Four locations were visited at St. Charles de Bourget (Fig. 4.2), three of exposed mineralization (2 Fe-Ti-V and 1 Fe-Ti-P) and one of which exposed some of the 1.16 Ga Lac Chabot ferrodiorite megadyke. A total of 18 samples were collected (full list given in Appendix 1) from which 10 were chosen for analytical work (massive oxides (n=5), host anorthosite (n=2), ferrodiorite (n=2), and nelsonite (n=1)). Sample locations are displayed in Fig. 4.2. Nelsonites from this drillcore in this area are already described by Martin-Tanguay (2012) and Fe-oxide and apatite data already exist for comparison (Dare et al. 2014; Desormiers 2015).

A large exposure of massive oxides (Fe-Ti-V) is exposed near the river in the southern portion of St. Charles de Bourget (samples MG-SCG-01-06). Here the massive oxides contain abundant xenoliths of anorthosite and dunite (Fig. 4.3A). Xenoliths of anorthosite are slightly rounded and locally very large, ranging in size from 10's of cm's to several meters in length. Dunite xenoliths consist of single coarse olivine crystals ranging from 2-5 cm in size or conglomerations of multiple crystals >10 cm in size. The groups of olivine crystals can locally contain interstitial Fe-oxides (Fig. 4.3B). The samples analyzed from this first outcrop were representative samples of massive oxide ore (MG-SCB-01a and MG-SCB-01b) and anorthosite from one of the large xenoliths (MG-SCB-03). Slightly to the east of the exposed massive oxides, the mylonitized contact with host anorthosite is exposed (Fig. 4.3C). One sample was taken from the massive oxide (MG-SCB-07) at the contact and another was taken from an anorthosite exposure nearby (MG-SCB-08).

A second large exposure of massive oxides is located near a cottage (MG-SCB-09-MG-SCB-13) approximately 500 m west of the first sampling area. The massive oxides also contain abundant large (m-scale) xenoliths of anorthosite and leucogabbro and cm-scale sized xenoliths of dunite (Fig. 4.3B). Disseminated Fe-oxides are present locally within the anorthosite/leuconorite xenoliths. Two samples of representative massive oxides were analysed: MG-SCB-11 represents typical massive oxide ore and MG-SCB-12 represents massive oxide ore containing fine-grained silicate xenoliths (olivine and plagioclase).

Nelsonite is exposed along the road roughly 1 km north of the massive oxide outcrops described above. Nelsonitic mineralization at St. Charles de Bourget is composed of massive nelsonite, apatite-rich (>50% locally) with abundant, locally coarse (>5 cm) xenoliths of dunite/single olivine crystals (Fig. 4.3D). Two samples of massive nelsonite (MG-SCB-14 and MG-SCB-16), representative of nelsonitic (Fe-Ti-P) mineralization from the St. Charles de Bourget area, were taken to supplement nelsonite samples from the Martin-Tanguay (2012) study of drillcore in that area.

The southern part of the Lac Chabot ferrodiorite megadyke is exposed as small outcrops along the road, approximately 2 km northwest from the nelsonite location. Two samples of ferrodiorite were taken of fine-grained dyke (MG-SCB-17) and medium-grained dyke (MG-SCB-18) containing approximately 5-10% Fe oxides either as mm-scale layers or disseminations (Fig. 4.3E). It has been proposed by Duchense and Liégeois (2015) that nelsonites form as cumulates from evolved ferrodioritic melts.

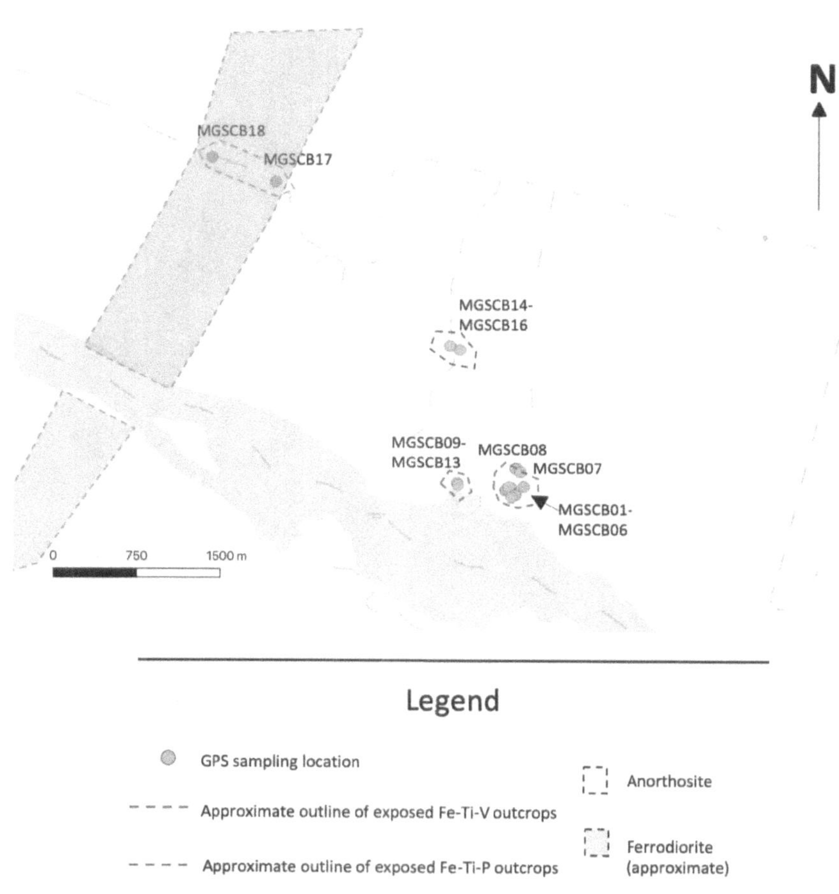


**Figure 4.2:** Map displaying the locations of samples taken from both the mineralized and area surrounding the St. Charles de Bourget area. Base layer of the map from ©2018 Google.

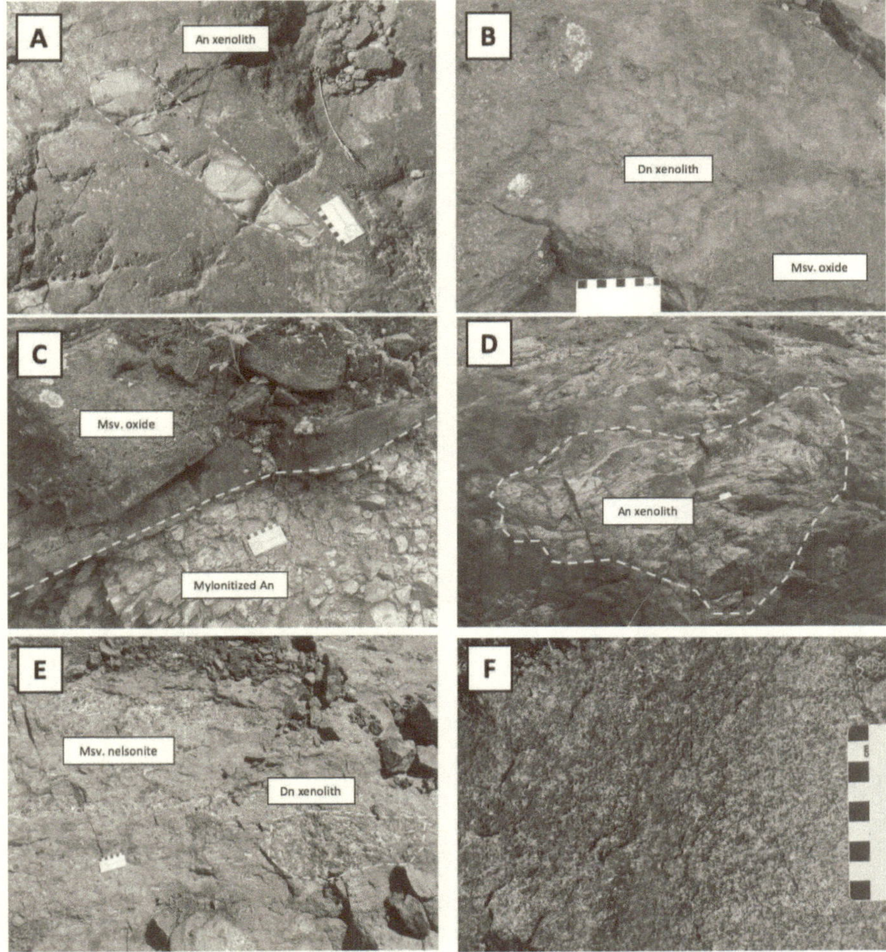


**Figure 4.3:** Outcrop photographs of mineralized zones (A-D) and associated silicate lithologies (E) within the St. Charles de Bourget area. A) Massive (Msv) oxide containing xenoliths of anorthosite (An) and B) dunite (Dun) from the first sampling area. C) Mylonitized contact between massive oxide body and host anorthosite from the first sampling area. D) Large anorthosite xenolith hosted within massive oxide body from the second sampling area. E) Massive nelsonite outcrop containing xenoliths of dunite. E) Medium-grained Lac Chabot ferrodiorite dyke, located to the northwest of the mineralized area.

4.1.2   Petrography

The petrographic study at St. Charles de Bourget involved examining a total of 8 polished thin sections (Appendix 1). In summary, both massive oxide (n=5) and nelsonite (n=1) lithologies are relatively medium-grained, magnetite-dominated, and display a wide variety of deformation-related textures which are described in this section. Host anorthosite (n=2) contains very few (if any) mafic minerals and appears to have undergone significant dynamic recrystallization based on hand sample observations. It is important to note that although both massive oxides and nelsonites in some cases contain silicates (10% Fe-rich olivine in particular: Martin-Tanguay 2012), they are still defined as massive based on the outcrop and hand-sample interpretation of these silicates to be small xenoliths. None of the ferrodiorite samples were examined petrographically as they are already described in Côté (1986).

The massive oxides at the St. Charles de Bourget deposit are magnetite-dominated as samples consist of approximately 55-70% magnetite, 20-30% ilmenite, and trace to 10% Al-spinel. The Fe-oxide phases are generally present as mm-sized crystals that have been subject to varying degrees of recrystallization as shown by triple junctions along magnetite grain boundaries (Fig. 4.4A). Exsolution phases are most prominent within magnetite crystals, and are present as (in order of most abundant exsolution phase); 1) Fine-grained cloth-textured spinel (Fig. 4.4B), 2) fine-grained linear trellis spinel and ilmenite (Fig. 4.4C), and 3) medium-grained (50-100 μm) sandwich exsolutions of ilmenite (rare). Ilmenite display a wide range in grain sizes (>100 μm to mm-scale) and are generally exsolution free. One interesting feature about these ilmenite crystals is the local presence of undulose extinction similar to that found in deformed quartz (Fig. 4.4D). This is most likely caused by solid state deformation of ilmenite exsolution granules. When present, spinel commonly form coarse, anhedral crystals that contain inclusions of magnetite or ilmenite.

Silicate minerals present within massive oxides consist of olivine and/or plagioclase which can be present as single crystals or groups of multiple crystals. The olivine crystals present are subhedral, slightly serpentinized, but relatively unaltered (Fig. 4.4E) while plagioclase has reacted significantly in the presence of Fe-oxides. This is shown by a 100-200 µm thick corona of amphibole ± biotite which surrounds plagioclase crystals in contact with Fe-oxide crystals (Fig. 4.4F). The coronas form likely due to the inability of plagioclase to incorporate Mg into its crystal structure during sub-solidus re-equilibration (Whitney and McClelland 1983).

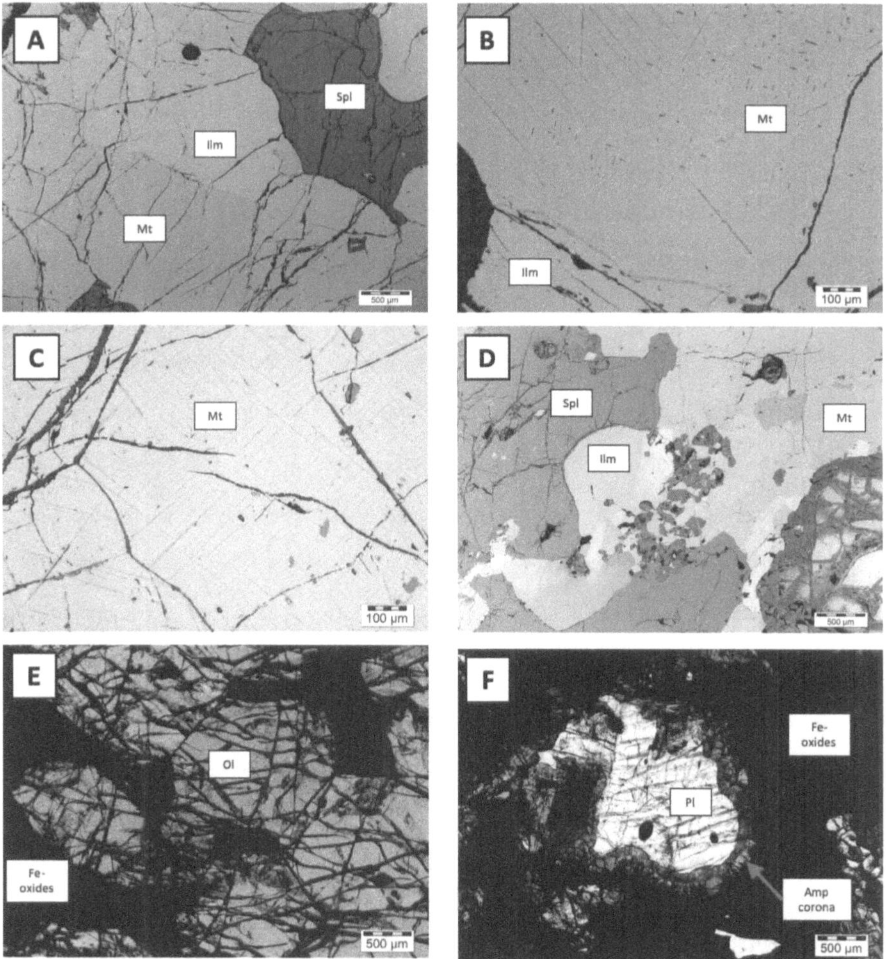


**Figure 4.4:** Photomicrographs of massive oxide samples from St. Charles de Bourget in reflected light (A-D) and transmitted light (E-F). A) coarse-grained magnetite (Mt), ilmenite (Ilm) and spinel (Spl) from sample MG-SCB-01a. B) fine-grained cloth-textured exsolutions of spinel within magnetite from sample MG-SCB-01b. C) Fine-grained trellis exsolutions of ilmenite and spinel within magnetite from sample MG-SCB-12. D) Ilmenite crystal displaying undulose anisotropy (cross-polars) within sample MG-SCP-07. E) clusters of olivine (Ol) from a dunite xenolith from sample MG-SCB-12. F) plagioclase (Pl) surrounded by a reaction corona of amphibole (Amp) when in contact with oxides (Ox) from sample MG-SCB-12.

Based on petrography performed by Martin-Tanguay (2012), nelsonite lenses contain, on average, 41% magnetite, 18% ilmenite, 29% apatite, and 10% olivine. The sole nelsonite sample from this study consists of approximately 30% magnetite, 10% ilmenite, 45% apatite, trace spinel, and 15% olivine (here interpreted as dunite xenoliths). Like in the massive oxides, Fe-oxides within nelsonites are relatively medium-grained (mm-sized crystals) and variably subjected to dynamic recrystallization but, magnetite lacks the diversity of internal exsolution textures (Fig 4.5A). The only exsolution texture observed is ultra-fine-grained, cloth-textured ilmenite exsolutions within magnetite (Fig. 4.5B), similar to that observed in massive oxide samples from Buttercup. Ilmenite is are exsolution free and can be present in a wide variety of grain sizes from >100 μm to several mm in length. When present, Al-spinel occurs as coarse-anhedral crystals. Apatite occurs as single subhedral to anhedral crystals approximately 300 μm to 1 mm in diameter or as clusters of multiple crystals (Fig. 4.5C). Some apatite crystals contain inclusions of Fe-oxides (Fig. 4.5D). Olivine xenocrysts/dunite xenoliths are present as very coarse-grained, subhedral-anhedral crystals which have been slightly serpentinized (Fig. 4.5E). It is significant to note that the unlike the massive oxides, the nelsonite lacks plagioclase (both in this study and that of Martin-Tanguay 2012). The nelsonite outcrop also generally lacks obvious anorthosite xenoliths.

Anorthosites associated with the St. Charles de Bourget deposit are composed of nearly 100% plagioclase with only trace amounts of Fe-oxides and biotite. Orthopyroxene was noted in trace amounts within anorthosite at St. Charles de Bourget by Martin-Tanguay (2012), however, it is not seen in this study. Anorthosite found near the contact of the massive oxide body contains plagioclase with well-developed granoblastic texture (Fig. 4.6C) indicating that the anorthosite underwent significant dynamic recrystallization and grain-size reduction (Fig. 4.6A). The anorthosite xenolith, hosted in the massive oxide body displays a protoclastic texture that consists of coarse plagioclase crystals surrounded by very fine-grained, recrystallized plagioclase. This texture can locally be moderately developed (Fig. 4.6B) or well developed (Fig 4.6C) within a single thin section. Twinning has practically been erased within the re-crystallized plagioclase but twinning within the coarse plagioclase crystals is deformed, consisting of tartan twins and discontinuous or bent albite twins.

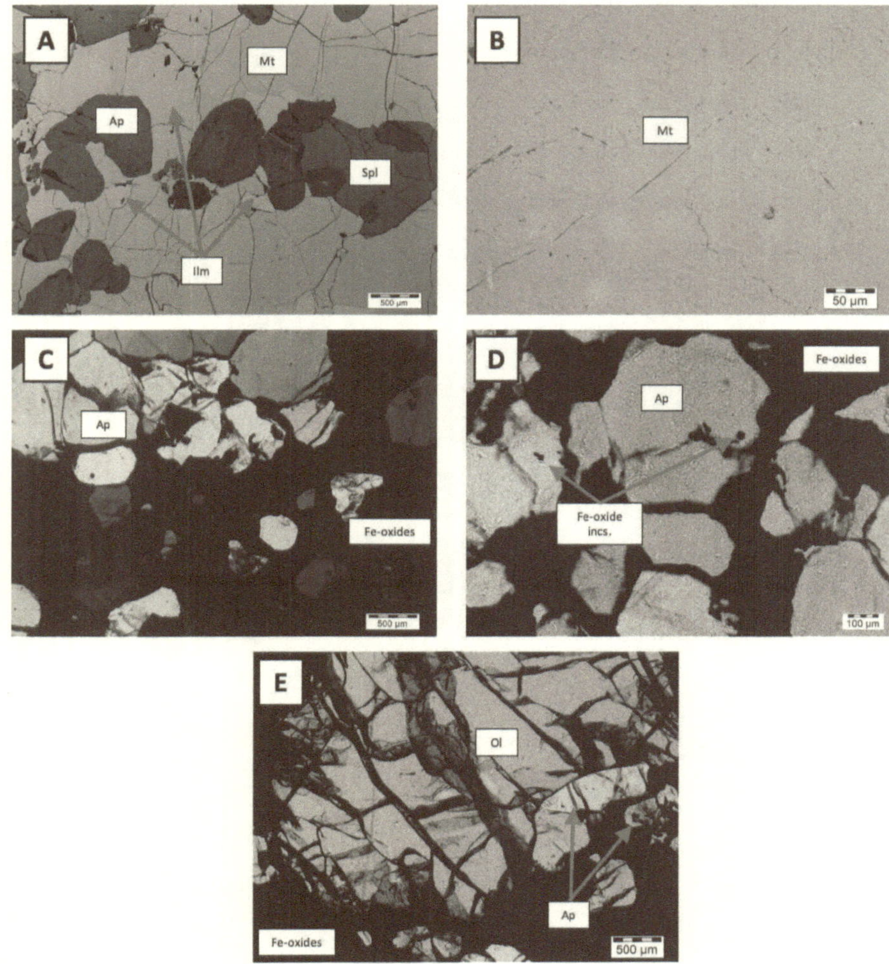


**Figure 4.5:** Photomicrographs of interesting features from a nelsonite sample (MG-SCB-14) in reflected (A-B) and transmitted (C-E) light. A) medium-grained magnetite (Mt), ilmenite (Ilm), and spinel (Spl). B) ultra-fine cloth-textured exsolutions of ilmenite within magnetite. C) apatite (Ap) crystal clusters and subhedral to anhedral apatite crystals. D) apatite crystals containing inclusions of Fe-oxides. E) olivine (Ol) from dunite xenolith.

56

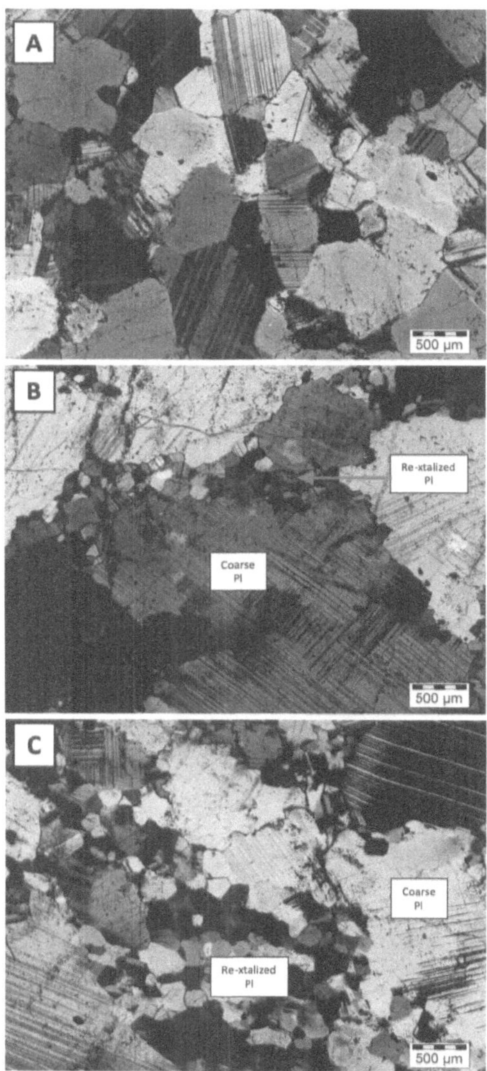


**Figure 4.6:** Representative photomicrographs of anorthosite surrounding mineralized areas (MG-SCB-08 inA) and from large xenoliths (MG-SCB-03 in B-C). A) plagioclase crystals displaying granoblastic texture with well-developed triple junctions indicating recrystallization. B) plagioclase crystals displaying moderately-developed protoclastic texture (grain size reduction between coarser, less deformed plagioclase crystals) and tartan twinning within the coarse crystals. C) plagioclase crystals displaying well-developed protoclastic texture.

## 4.2    Lac Perron

The Lac Perron Fe-Ti-P occurrence is located along the western margin of the Lac St. Jean anorthosite suite (Fig. 2.6). This occurrence is at the prospection stage and previous work is limited to claims reporting (Tremblay 2014). Mineralization within the area was first discovered in 2002 by prospector (Reynald Desgangé) who noted oxide-rich rocks along a forestry road. Outcrop stripping and geochemical sampling (Tremblay 2014) indicates that 1) the mineralization consists of 10's of m-scale lenses of massive oxides and apatite (nelsonite), containing 0.5-22 % $P_2O_5$ (Table 2.1) and 2) the deposit is hosted within very coarse-grained anorthosite with localized patches of mafic minerals.

### 4.2.1    Sampling and outcrop descriptions

Sampling at the Lac Perron occurrence was performed over a relatively small area ($<500m^2$) due to limited exposure of the mineralization. Exposed massive nelsonite (Fe-oxide and apatite) consists of an approximately 100 m long NW-SE-trending lense exposed along the forestry road. A total of 11 samples were taken from this nelsonite body (Fig. 4.8). Outcrops of anorthosite are exposed along the same road approximately 300 m south of the nelsonite exposure, where one sample was taken, and also a few 100 m's north.

The massive nelsonite lenses are composed of very coarse-grained apatite and Fe-oxides and display sharp but, irregular contacts with the host anorthosite (Fig. 4.9A). Along the contacts between anorthosite and nelsonite, the nelsonite appears to display a moderately developed lineation defined by elongated apatite crystals. Samples of anorthosite (MG-LPR-4a) and nelsonite (MG-LPR-4b) were taken at this contact. Away from the contact, nelsonite appears more massive and is not lineated; the apatite crystals become very coarse-grained (>3cm), irregular-shaped crystals and can contain abundant inclusions of Fe-oxides (Fig. 4.9B). Due to the coarse-grain size of apatite, sampling is heterogeneous: sample MG-LPR-5

represents apatite-poor (<20% apatite) portion and MG-LPR-6 representing apatite-rich
(>50% apatite) portion of the nelsonite. An additional 2 samples (LPR-1-2), provided ahead of
the field visit, includes an apatite-free massive oxide sample from the same outcrop. In the
southern portion of the nelsonite lense, massive nelsonite grades into semi-massive nelsonite
(MG-LPR-8 and MG-LPR-10) towards the east (Fig. 4.9C) and anorthosite containing trace
Fe-oxides (MG-LPR-7 and MG-LPR-9) towards the west (Fig. 4.9D). In both the semi-
massive nelsonite and anorthosite, Fe-oxides in contact with plagioclase are locally
surrounded by a corona of either amphibole or garnet.

The anorthosite sample (MG-LPR-11), taken south of the nelsonite exposure,
represents the host anorthosite because it distant from the mineralization and barren of any Fe-
oxides and apatite. Plagioclase crystals from this outcrop are very coarse-grained and
weathered white (Fig. 4.9E). Outcrops of anorthosite and leuconorite are exposed to the north
of the nelsonite ore body, and locally contain megacrystic orthopyroxene crystals (Fig. 4.9F.).
However, these megacrysts likely do not contain any relict orthopyroxene, as it has been
almost completely converted to biotite.

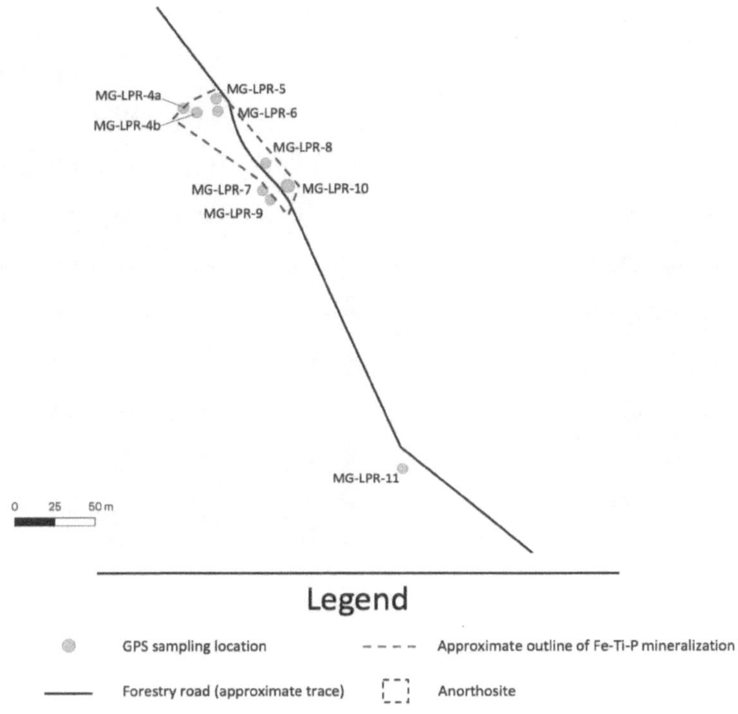


**Figure 4.7:** Map displaying the locations of samples taken from both the mineralization and surrounding anorthosite at the Lac Perron occurrence. Base layer of the map from ©2018 Google.

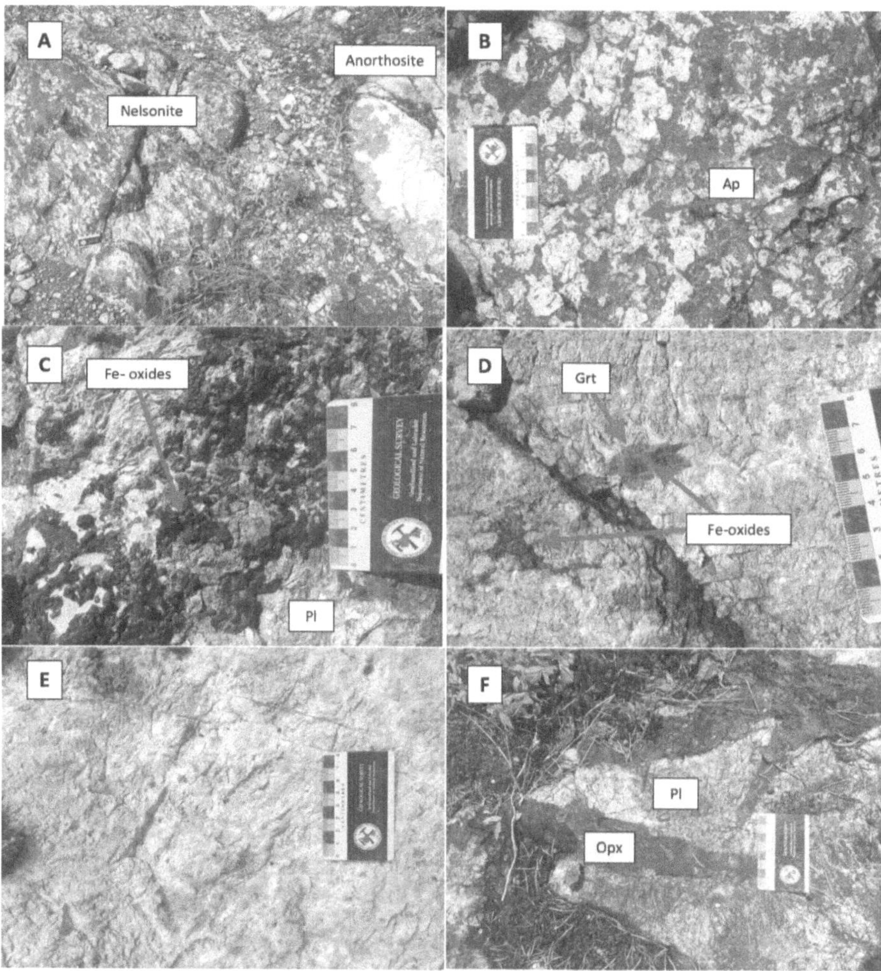


**Figure 4.8**: Outcrop photographs of the mineralization and host rocks from Lac Perron. A) Contact between massive nelsonite and host anorthosite. B) Apatite-rich nelsonite, note the Fe-oxides can locally be included within very coarse-grained apatite (Ap) crystals. C) Semi-massive/disseminated nelsonite, note the presence of amphibole coronas locally around Fe-oxides when in contact with plagioclase (Pl). D) Patches of Fe-oxides within anorthosite adjacent to the mineralization. Note the local presence of garnet (Grt) coronas surrounding Fe-oxides. E) Anorthosite host located south of the mineralization. F) Leuconorite with orthopyroxene (Opx) megacrysts located north of the mineralization.

### 4.2.2    Petrography

The nelsonites at Lac Perron are composed of approximately 30-45% magnetite, 25-30% ilmenite, and about 5-10% spinel (Fig. 4.10A). Due to the heterogeneous distribution of very coarse-grained apatite within the mineralized body, modal proportions of apatite range from 0-45%. The Fe-oxides have been variably subjected to dynamic recrystallization within the centre of the mineralized body and display elongation near the contact with anorthosite. Both magnetite and Al-spinel can contain exsolutions but, the most diverse exsolution textures are displayed within magnetite. Magnetite contains both fine-grained cloth-textured spinel exsolutions (Fig 4.10B) and coarse sandwich textured exsolutions of ilmenite (Fig. 4.10C): both sets of exsolutions are oriented relatively 90 degrees to each other. Al-spinel contains very fine-grained blebby exsolutions of magnetite which are concentrated towards the core of the crystal (Fig 4.10D). Apatite crystals within the nelsonites consist of either singular crystals or groups of multiple coarse-grained crystals (Fig 4.10E). Apatites appear to engulf Fe-oxides locally and contain abundant sub-micron inclusions of Fe-oxides which can be arranged in a discontinuous linear pattern within a single crystal (Fig. 4.10F). Within the semi-massive nelsonite, several mm-thick coronas of amphibole + biotite surround the Fe-oxides when they are in contact with plagioclase (Fig 4.10G).

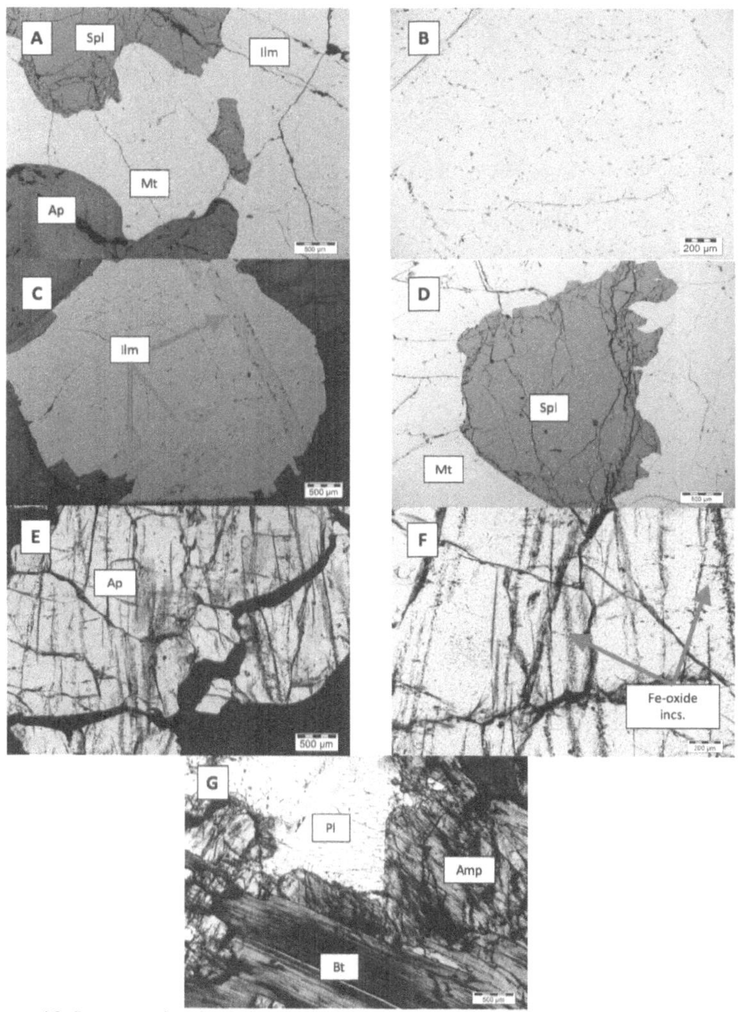

**Figure 4.9:** Representative photomicrographs from the mineralization within Lac Perron. A) coarse-grained magnetite (Mt), ilmenite (ilm), spinel (Spl) and apatite (Ap) within massive nelsonite (sample.MG-LPR-6). B) cloth-textured spinel exsolutions within magnetite from sample MG-LPR-04b. C) sandwich-textured ilmenite exsolutions within magnetite crystals within sample MG-LPR-04b). D) coarse spinel crystal displaying fine-grained exsolutions of Fe-oxide from sample MG-LPR-04b. E) coarse-grained apatite crystals within nelsonite from sample MG-LPR-04b. F) apatite containing micro inclusion trails of Fe-oxides from sample MG-LPR-04b. G) corona of amphibole (Amp) and biotite (Bt) at the contact between plagioclase (Pl) and Fe-oxides from sample MG-LPR-08.

Anorthosites at Lac Perron are composed of 95-100% coarse-grained (>2cm) plagioclase (Fig. 4.11A). Plagioclase crystals are generally anhedral to subhedral and display sutured grain boundaries with other plagioclase grains (Fig. 4.11B). Moderate to strong, pervasive serecite alteration of single plagioclase crystals (Fig. 4.11C) and along grain boundaries is common within anorthosite samples close to (MG-LPR-07) and further from (MG-LPR-11) the mineralization at Lac Perron. Fe-oxide minerals are only found locally within anorthosite close to mineralization, but in very low modal percentages (0-5%). Present Fe-oxide minerals are usually surrounded by a >0.5 cm wide corona composed of a biotite and amphibole inner layer with a garnet outer layer (Fig. 4.11D).

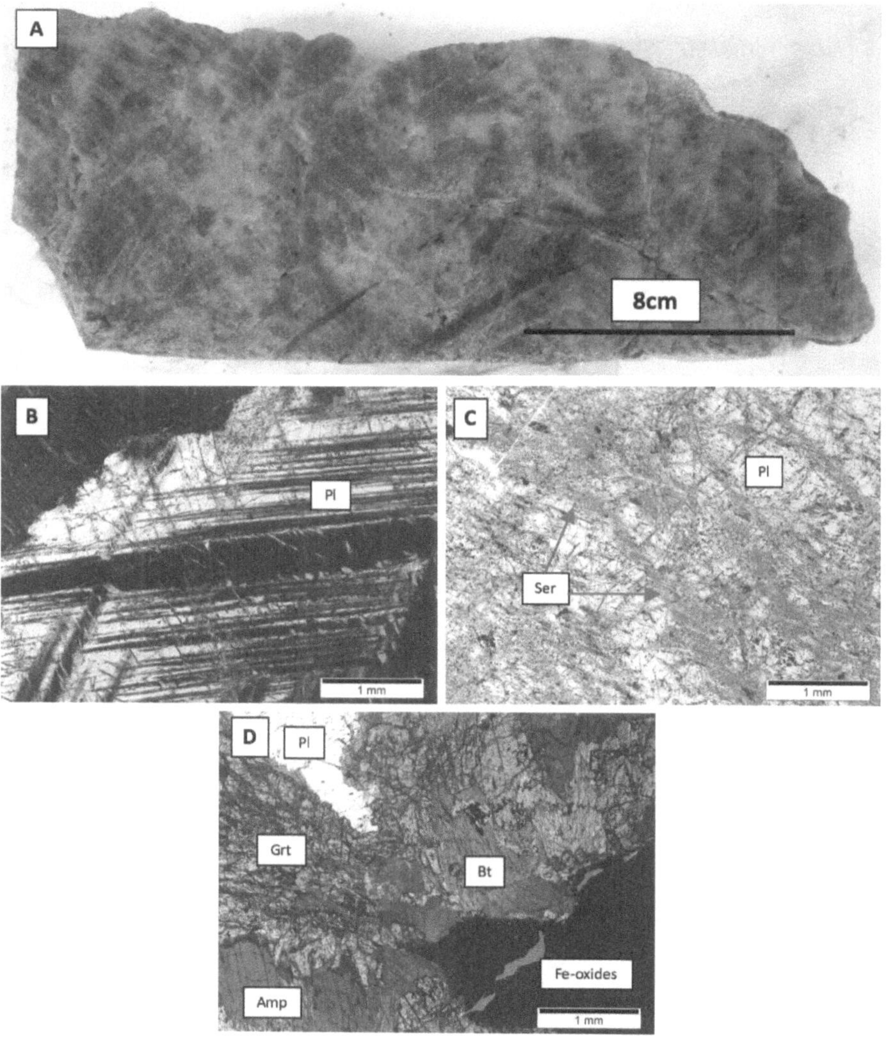



**Figure 4.10**: Photographs (A) and photomicrographs (B-D) of important textures from anorthosite at Lac Perron. A) sample photograph of MG-LPR-11. B) Plagioclase (Pl) crystals with sutured grain boundary from sample MG-LPR-11. C) strong sericite (Ser) alteration of a plagioclase crystal from sample MG-LPR-11. D) amphibole (Amp), biotite (Bt), and garnet (Grt) coronas surrounding Fe-oxide crystal from sample MG-LPR-08.

65

## 4.3    Lac à Paul

The Lac à Paul Fe-Ti-P deposit is located within the northern-most region of the Lac

St. Jean anorthosite suite, approximately 40-60 km east of the Lac Margane/Houlière V

showings. The Lac-à Paul region contains 9 different nelsonite zones (Manouane, Paul,

Nouvelle, Nicole, Lucie, Lise, Intersection, Castor, No. 1, and No. 2 see Fig. 4.11). The

largest mineralised zone at Lac à Paul (Paul Zone) is approximately 2700 m in length and

150-300 m wide (Arianne Phosphate Inc. 2013). The calculated grade and tonnage are

reported in Table 2.1.

Mineralization within the Lac à Paul region onsists of olivine nelsonite and nelsonitic

gabbro/norite hosted within a sequence of mafic to ultramafic rocks, such as anorthosite,

leuconorite, norite, gabbronorite, olivine gabbro, pyroxenite, and locally peridotite, dunite,

and magnetitite (Arianne Phosphate Inc. 2013). Ferrodiorite dykes have been noted to the west

of the Lac à Paul claim area (Fig. 4.11; Arianne Phosphate Inc. 2013). These lithological

facies are layered in a disorderly fashion and lithological contacts are more or less diffuse

(Arianne Phosphate Inc. 2013). The petrography of the mineralized zones is described in

Chartier-Montreuil et al. (2017) as follows; the high-grade ore forms the northern-most layer

and comprises fine-grained nelsonitic peridotites containing 20-40 %, each, of olivine, Fe-

oxides (mainly ilmenite) and apatite. The low-grade ore forms the southern-most layer and is a

coarse-megacrystic nelsonitic olivine norite (termed apatite gabbro in Fig. 4.11). Trace

element data of apatite from the ore zones of Lac à Paul are taken from the study of Chartier-

Montreuil et al. (2017) to compare with apatite analysed in this study.

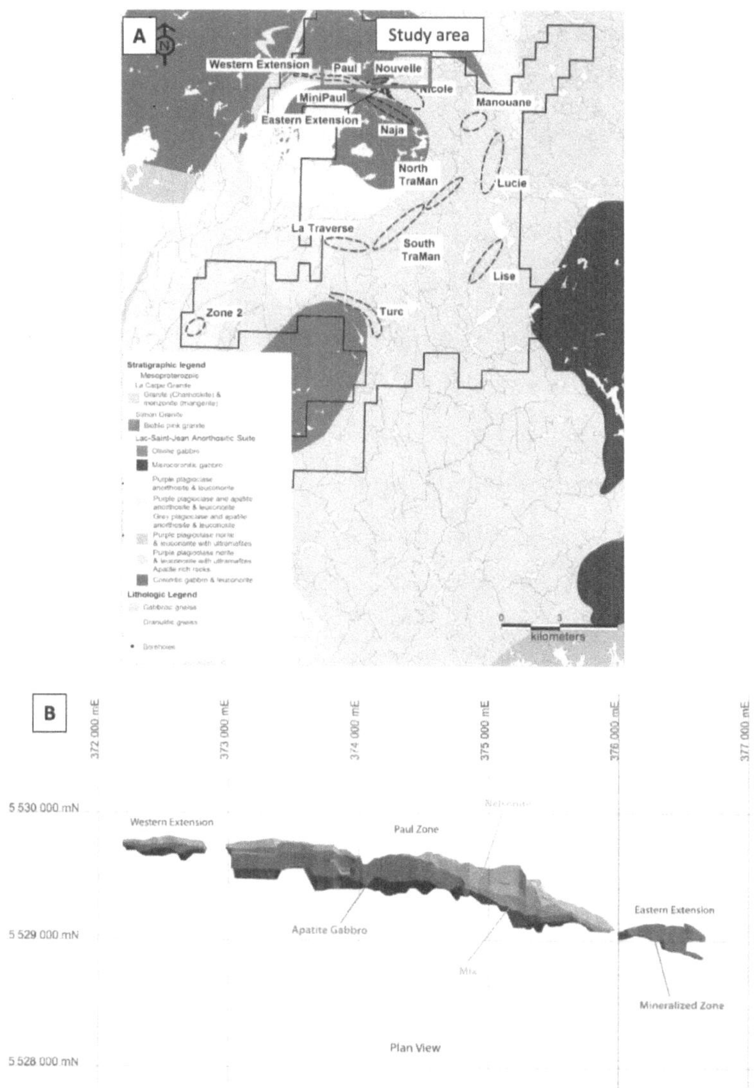


**Figure 4.11:** Bedrock geology and mineralization within the Lac à Paul area modified from Arianne Phosphate Inc. (2015). A) Claim boundary map with names and locations of all mineralized zones. B) Cross-section through the Paul Zone.

67

4.3.1  Sample descriptions

The 4 samples used in this project are from the oxide-rich gabbroic footwall rocks (termed Zone Nouvelle, i.e. New Zone) to the northeast of the Paul Zone (Fig. 4.11A). The samples were first described by Kobylinski (2015) and chosen for this project due to elevated modal percentages of Fe-oxides and apatite. The New Zone show a high magnetite anomaly and is unusually rich in Fe-oxide and apatite compared to the rest of the footwall gabbronorites. The New Zone is characterised by alternating layers of oxide-rich to oxide-poor gabbro/norites and samples were taken across this stratigraphy (S. Dare pers. Comm.). It is overall much lower in apatite than the ore zones but locally still contains some apatite, which were analysed in this study. The 4 samples comprise; 1) dunitic nelsonite (SDLP-18), 2) nelsonitic olivine norite (SDLP-36a and b), 3) nelsonitic pyroxenite (SDLP-37), and 4) nelsonitic norite (SDLP-38). All samples are generally medium-grained and contain significant proportions of silicate minerals in order to classify them as semi-massive (>30% Fe-oxides) or disseminated (<30% Fe-oxides). Dunitic nelsonites and nelsonitic olivine gabbronorites are considered semi-massive nelsonite while nelsonitic pyroxenites and nelsonitic norites as disseminated nelsonites (see Appendix 1).

4.3.2  Petrography

Significant differences between Fe-oxides and silicates within semi-massive and disseminated samples are also noted on the thin section-scale. The Fe-oxide mineral textures vary with 1) modal percentages of Fe-oxide phases (magnetite, ilmenite, Al-spinel), 2) grain size, and 3) quantity of exsolutions (exsolutions are only present in magnetite). Within the semi-massive oxide-rich samples, magnetite is usually the dominant Fe-oxide phase; magnetite ranges from 20-25 %, ilmenite ranges from 10-15 %, Al-spinel is present from trace amounts to 5%, and apatite modal percentage ranges from 1-5 %. The dunitic nelsonite sample (SDLP-18) contains only olivine while the nelsonitic olivine norite sample contains plagioclase and orthopyroxene, with minor olivine as primary silicate phases. Fe-oxides within the semi-massive samples are generally >500 μm in size, display varying degrees of dynamic recrystallization, and contain distinct grain boundary textures between the different phases (Fig 4.12A). When magnetite and ilmenite grains are in contact with each other, ilmenite grain boundaries appear partially serrated, and contain a partial corona of fine-grained spinel granules. Magnetite from semi-massive samples is generally exsolution-rich as it contains abundant trellis exsolutions of ilmenite and spinel (Fig. 4.12A).

Within the disseminated oxide-rich samples, Fe-oxides are generally fine-grained (Fig 4.12B and C) and ilmenite is the dominant phase (2.5-10 %) while that of magnetite is subordinate (2.5-5 %), and Al-spinel is rare to absent within these samples. Apatite generally comprises trace amounts to 1 % of these samples. Ilmenite crystals are also significantly coarser-grained than magnetite crystals (Fig. 4.12B). Fe-oxide phases generally display stronger deformation-related textures such as a general foliation and elongation of most crystals (Fig. 4.12C). Fine-grained symplectites of Fe-oxides are also locally present within these samples (Fig. 4.12B). Exsolutions are relatively absent within all Fe-oxide phases from these samples.

Within the semi-massive samples, primary silicate phases consist mostly of olivine and orthopyroxene while the disseminated samples contain orthopyroxene, clinopyroxene, and plagioclase with secondary hornblende and biotite. Olivine within the semi-massive samples is generally unserpentinized, coarse-grained and surrounded by a corona of orthopyroxene ± symplectic Fe-oxides (Fig. 4.12D). From semi-massive to disseminated samples, orthopyroxene and clinopyroxene transition from coarse-grained crystals locally containing exsolutions of plagioclase and ilmenite to skeletal crystals which can locally display significant alteration to hornblende ± biotite (Fig. 4.12E). Plagioclase is only present in samples with disseminated samples and displays well-developed granoblastic texture (Fig. 4.12F) and twin boundary migration recrystallization. Apatite shows little change from semi-massive to disseminated samples although, it is generally finer-grained within disseminated samples.

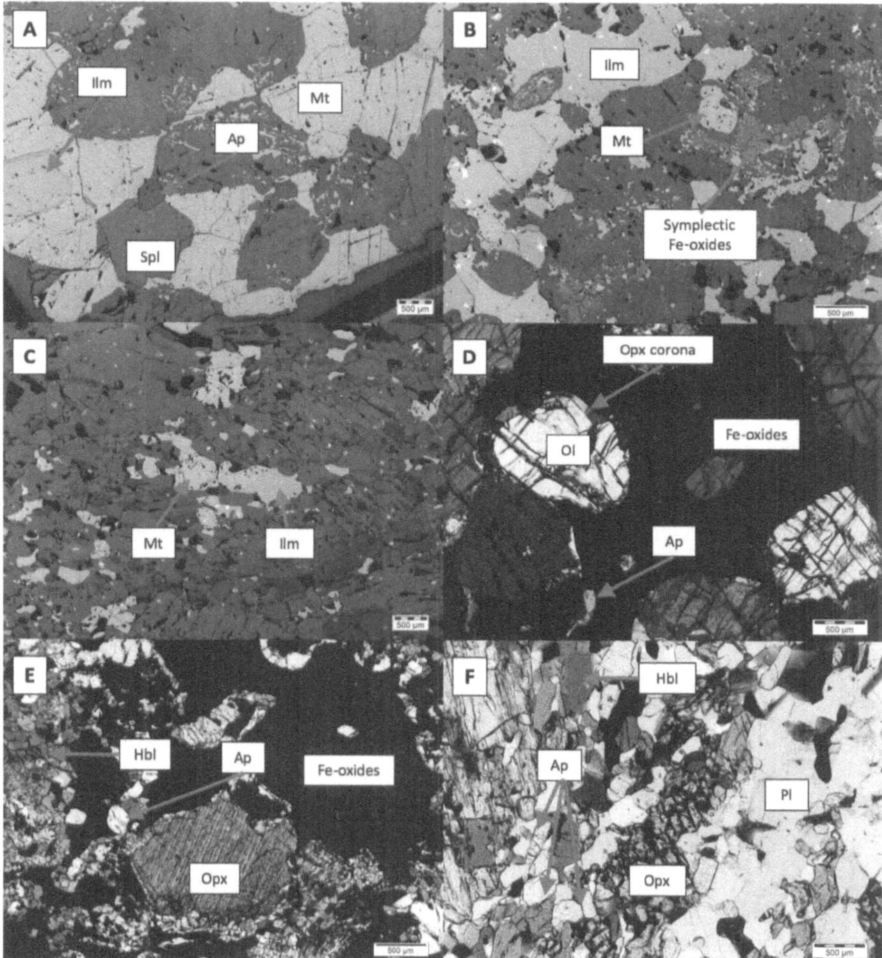


**Figure 4.12:** Reflected (A-C) and transmitted light (D-F) photomicrographs of ore-bearing samples from the Lac à Paul deposit. A) semi-massive oxides composted of magnetite (Mt), ilmenite (Ilm), and spinel (Spl) within the the dunitic nelsonite (sample SDLP-18). B) disseminated Fe-oxides from the nelsonitic pyroxenite (SDLP-37), oxides from the nelsononitic olivine gabbronorite display similar textures. C) disseminated Fe-oxides from the nelsonitic norite (sample SDLP-38). D) olivine (Ol) crystals and fine-grained apatite (Ap) from the dunitic nelsonite (SDLP-18), note the thin orthopyroxene coronas surrounding the olivine. E) orthopyroxene (Opx) and secondary hornblende (Hbl) from the nelsonitic pyroxenite (SDLP-37). F) plagioclase (Pl), orthopyroxene, hornblende, and apatite from nelsonitic norite (SDLP-38).

71

4.4    Lac à l'Orignal

The Lac à l'Orignal showing is hosted within the 1080 Ma Lac Vanel anorthosite and as such is quite different from the other Fe-oxide-apatite deposits/showings from the 1170-1140 Ma Ga Lac St. Jean anorthosite described above. The mineralization at Lac à l'Orignal is composed of a ferrogabbro unit, approximately 50-70 m thick (Fig. 4.14) with approximately 25 % Fe-oxides and 8-15% apatite, hosted within an anorthosite unit with local patches of leuconorite and syenite/monzonite (Fig. 4.13; Glen Eagle Resources Inc. 2017). Recent drilling of the Lac à l'Orignal showing gives an average grade of 5.1 % $P_2O_5$, with grades locally of up to 7 % $P_2O_5$ Glen Eagle Resources Inc. (2017).

4.4.1    Sample descriptions and petrography

Samples of nelsonitic norite (n=3), anorthosite (n=1), and massive oxide (n=1), analyzed during this project, originated from drill core obtained from Glen Eagle resources Inc. The drill hole locations are displayed in (Fig. 4.13). Samples of nelsonitic norite were taken from drill holes LO12-10 at a depth of 51.3 m (MG-LO-01) and LO12-31 at a depth of 93 m (MG-LO-03). The drill hole, depth, and location of the sample coded LacOrignal is unknown. Nelsonitic norite samples are generally medium-grained, moderately foliated, and contain approximately 10-20 % Fe-oxides, 10 % apatite, with 40 % plagioclase and 40 % orthopyroxene (Fig 4.14D). The sample of massive oxide (MG-LO-04) was obtained from drill hole LO12-13 at a depth of 90.6 m which stratigraphically underlies the main nelsonitic norite unit. Massive oxides are coarser-grained relative to the nelsonitic norites, generally have equal proportions of magnetite (40 %) to ilmenite (40 %) along with 10 % spinel, 5 % sulfide minerals, and 5 % orthopyroxene. The sample of anorthosite (MG-LO-02), obtained from drill hole LO12-25 at a depth of 3.9 m-stratigraphically above the nelsonitic norite unit, displays a well-developed granoblastic texture, composed of <90% plagioclase with local

patches of orthopyroxene.

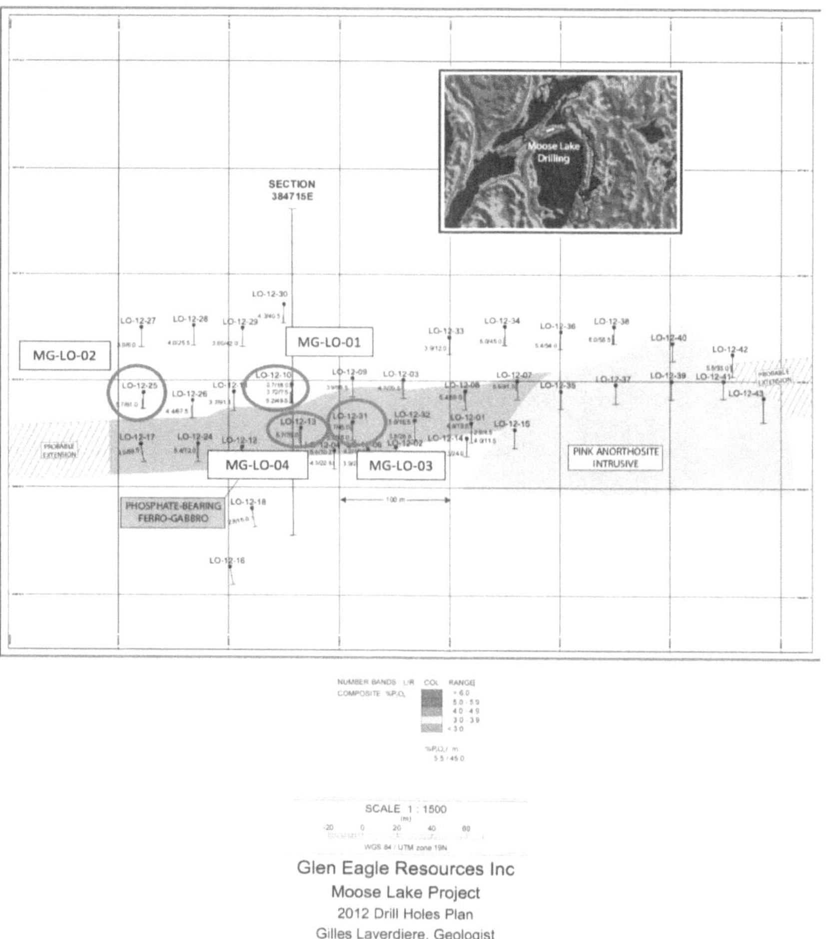



**Figure 4.13:** Drilling plan from the 2012 exploration of the Lac à l'Orignal occurrence, the locations of the drill holes along with the sample code for those samples used for this project are annotated. Figure modified from Glen Eagle Resources Inc. (2017).

4.4.2   Petrography

When comparing the mineralization from Lac à l'Orignal to other deposits examined during this project, there are two significant differences; 1) this is the only mineralization where hemo-ilmenite is observed and 2) sulfides are much more common (~5 %) within the oxide-apatite mineralization than among the other deposits/showings (trace amounts). This section will explain the differences from the other deposits and any textural differences within the thin sections examined.

Nelsonitic norites from the ferrogabbro unit are plagioclase- and orthopyroxene-rich rocks which contain approximately 5-10 % hemo-ilmenite, trace-5 % magnetite, 5-15 % apatite, with trace Al-spinel and sulfide minerals. Hemo-ilmenite within the nelsonitic norite contains abundant coarse-grained, tabular exsolutions of hematite extending across the length of the host crystal (Fig. 4.14A). These exsolutions decrease in size when the host ilmenite crystal is in contact with a magnetite crystal (Fig 4.14B). Magnetite within the nelsonitic norites generally appears to lack significant exsolutions. Apatite is generally in close association with the Fe-oxides and occurs mostly as clusters of subhedral crystals with similar grain size to the Fe-oxides (Fig. 4.14C).

The massive oxide sample contains 45 % magnetite, 35 % ilmenite (not hemo-ilmenite), 5 % Al-spinel, and 5 % sulfides. Arguably, the most notable of the differences between Fe-oxide textures within the massive oxide and the nelsonitic norites are with respect to exsolution textures. Ilmenite appears free of exsolutions while magnetite from the massive oxide can contain significant exsolutions of granular Al-spinel and/or tabular ilmenite crystals (Fig 4.14D). Sulfide minerals occurring within the massive oxide sample consist of mostly coarse-grained pyrrhotite crystals containing finer-grained pentlandite and chalcopyrite (Fig 4.14D).

Primary silicate phases within the nelsonitic norites have undergone significant deformation but do not appear to have been affected by corona reactions. Deformation is shown most significantly within plagioclase as crystals have been strongly subjected to dynamic recrystallization. Deformation of albite twins is shown by pinch and swell textures, bending, and the formation of tartan twins indicates that twin boundary migration recrystallization occurred. Orthopyroxene crystals exist in various different sizes: within sample MG-LO-03 they are a similar size to the other co-existing mineral phases but in sample MG-LO-01, they are very coarse-grained (locally several cm). Within both of these samples, orthopyroxene can contain exsolutions of ilmenite or rutile (Fig. 4.14E). Biotite only occurs locally as a secondary phase related to corona reactions that is closely associated with the contacts between Fe-oxides and plagioclase.

The host anorthosite was prepared into two thin sections, due to the heterogeneity of orthopyroxene within the piece of drill core that was used to prepare the sample: Sample MG-LO-02a contains 100 % plagioclase while MG-LO-02b contains 60 % plagioclase and 40 % orthopyroxene. Both samples contain hemo-ilmenite and biotite as trace mineral phases. Overall these samples display well developed granoblastic texture due to recrystallization where plagioclase displays serrated boundaries and twin boundary migration recrystallization (Fig. 4.15A). Orthopyroxene crystals are often grouped together possibly due to the development of an immature foliation (Fig 4.15B). Unlike within the mineralized samples, these orthopyroxene crystals within the anorthosites do not contain visible exsolution phases. Granoblastic texture and grain size reduction is also present locally within these crystals.

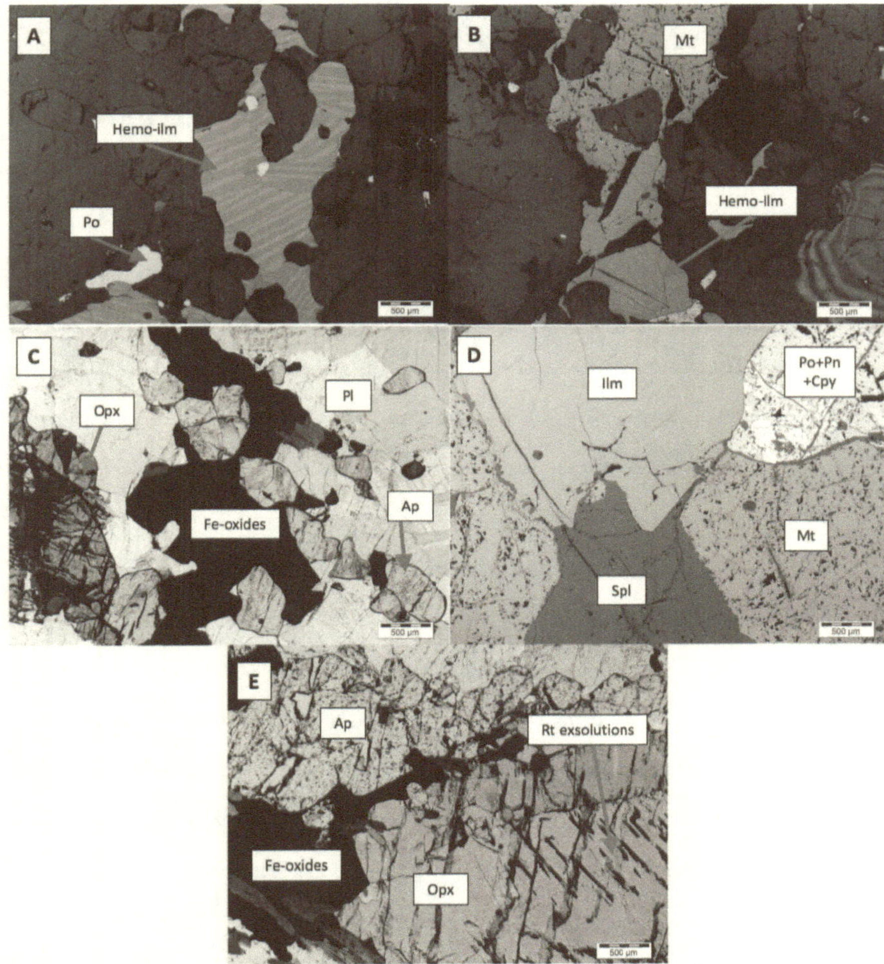


Figure 4.14: Reflected (A-B) and transmitted light (D and E) photomicrographs displaying key textures seen within the ore-bearing lithologies from the Lac à l'Orignal occurrence. A) Hemo-ilmenite (Hemo-ilm) and trace sulfides (Sul) from within nelsonitic norites (sample LacOrignal). B) Magnetite (Mt) and hemo-ilmenite grain contacts within nelsonitic norites (sample MG-LO-01), note the decrease in size of the hematite exsolutions. C) plane-polarized image of nelsonitic norite showing apatite and Fe-oxide mineralization intermixed with plagioclase and orthopyroxene (sample MG-LO-01). D) magnetite, ilmenite, Al-spinel (Spl) along with pyrrhotite (Po), Pentlandite (Pn), and chalcopyrite (Cpy) within massive oxide (sample MG-LO-04), note the absence of exsolutions within ilmenite. E) apatite and pyroxene from the nelsonitic norite, note the presence of exsolutions within orthopyroxene (sample MG-LO-03).

**Figure 4.15:** Photomicrographs of the host anorthosite from the Lac à l'Orignal occurrence. A) cross-polarized image of plagioclase displaying granoblastic texture and twin boundary migration recrystallization (sample MG-LO-02a). B) leuconorite-rich area within the host anorthosite (sample MG-LO-02b), note the grouping of fine-grained orthopyroxene crystals.

# Chapter 5  EMPA results

Electron microprobe analysis (EMPA) was performed to determine the major and minor composition of Fe-oxide minerals (magnetite, ilmenite, Al-spinel), apatite, and silicate minerals. Since the trace element content of Fe-oxides and apatite were also analyzed by LA-ICP-MS, EMPA data was used as the internal standard value during the data reduction process (total Fe for Fe-oxides and Ca for apatite; see Chapters 6 and 7). The results also provide initial insights on 1) Fe speciation and stoichiometry in magnetite and ilmenite, 2) composition of Al-spinel, 3) composition and classification of apatite based on volatile content (F-Cl-OH), and 4) classification and composition of the silicate minerals (e.g., An content of plagioclase, Mg# of pyroxene and olivine, type of amphibole, biotite and garnet). Overall, the results obtained by EMPA of all mineral phases complements the more detailed analyses of Fe-oxides and apatite obtained by LA-ICP-MS in the following Chapters.

## 5.1  Methods

EMPA analysis was performed at the University of Ottawa's microanalytical laboratory using a JEOL JXA-8230 superprobe fitted with 5 WDS (wavelength dispersive spectrometers). The analytical protocol for the analysis of Fe-oxides, apatite and silicates analyses differ and are thus presented separately. Full EMPA results for Fe-oxides are presented in Appendix 2, apatite results are presented in Appendix 3 and silicate results are presented in Appendix 4.

### 5.1.1  Fe-oxides

The Fe-oxide minerals analyzed by EMPA in this project were magnetite (n=383), ilmenite (n=254), and Al-spinel (n=171). Eight–12 grains were analysed per mineral per thin section. Concentrations of Al, Ca, Cr, total Fe, Mn, Mg, Ni, Si, Ti, Zn were measured as weight percent oxide (wt.%). Iron speciation was calculated by separating total FeO into FeO and $Fe_2O_3$ by stoichiometric charge balance. Analytical settings for EMPA analyses involved using an electron beam with a diameter of 20 μm, a current of 40 nA, and an energy of 20 kV. Using a beam with this diameter will ensure the incorporation of micro-exsolution phases into the analysis and the high beam current was used to ensure reasonably low detection limits of ~100 ppm for each element analyzed. The analytical protocol at the University of Ottawa microanalytical lab for the calibration and data treatment of Fe-oxides, including correction of interferences of Ti on V, and V on Cr, follows that first described in Duparc et al. (2016). Care was taken to make sure the EMPA beam avoided incorporating coarse exsolutions (>50 μm) within the respective Fe-oxide phases, when present. Reference materials were analysed as unknowns to verify the calibration, including the in-house standard BC-28 (magnetite from the Bushveld Complex: Dare et al. 2014). Data quality interpretations and comparison of the EMPA and LA-ICP-MS results for BC-28 are presented in Chapter 6. Analyses with totals higher than 101.5 wt.% and lower than 98.5 wt.% were not considered valid (i.e., poor quality analysis) and excluded from the sample average.

5.1.2   Apatite

Around 8-17 grains were analysed per thin section for a total of 127 grains of apatite from 9 samples. Analytical settings of the EMPA involved using a beam diameter of 10 μm, a current of 10 nA, and an energy of 15 kV. The beam energy was lowered and the beam size increased in order to more accurately determining the concentration of volatile elements (F, Cl), following the protocol of Marks et al. (2012) and Desormiers (2015), by negating the effect of anisotropic diffusion of F during analysis. This led to the detection limits (Table 7.4) being much higher than normal for minor/trace elements, such as the REEs, within apatite. This is not a problem as the trace elements are best determined by laser ablation ICP-MS. The list of standards used for calibration of apatite at the University of Ottawa is given in Desormiers (2015). In-house reference material Durango apatite (apatite from the Cerro de Mercado Iron-oxide apatite (IOA) deposit near Durango City, Mexico) was analyzed at the beginning of each session to monitor data quality (see appendix C). Formula units were calculated for each element in apatite assuming 12 oxygen atoms. Oxygen was calculated by cation stoichiometry and included in the matrix correction. Oxygen equivalent from halogens (F, Cl) was subtracted in the matrix correction. The OH concentrations were not measured by EMPA itself but were obtained through stoichiometric calculation $(XOH = 1 - (XF + XCl))$ where $X$ represents the formula units for each anion.

### 5.1.3   Silicates

Silicate minerals analyzed by EMPA include plagioclase (n=210), olivine (n=49), orthopyroxene (n=70), amphibole (n=93), biotite (n=101), and garnet (n=26). When present, around 4-8 points were analysed per mineral per thin section; but the number of grains analyzed in each sample was highly dependent on the modal percentage of the respective silicate phase in that sample. Nepheline was surprisingly identified in one sample from Buttercup deposit (BCP-2) as it was originally misidentified as plagioclase during petrography due to very-fine grain size and symplectic nature of the grains. The element list for each silicate phase analysed comprises Al, Ca, Cr, Fe, K, Mn, Mg, Na, Ni, Si, and Ti. Analysis of the hydrous mineral phases, such as amphibole and biotite, used the same element list but with the addition of F and Cl. Analytical settings were normal and the same for all silicate minerals: the beam diameter was set to 5 μm, beam current was 20 nA, and the beam energy was 20 keV. For anhydrous silicates (plagioclase, orthopyroxene, olivine, and garnet), totals higher than 101.5 wt.% and lower than 98.5 wt.% were not considered valid. For hydrous silicates, such as amphibole and biotite, the range of acceptable totals are different due to the fact that EMPA analysis is unable to detect OH. For amphibole, totals higher than 97.5 wt.% and lower than 95 wt.% were not considered valid and for biotite totals higher than 96.5 wt.% and lower than 93 wt.% were considered invalid.

## 5.2 Major and minor element composition of Fe-oxides

Magnetite can be categorized into two distinct groups based on petrography and average $TiO_2$ contents: high-Ti (>2.5 wt.% $TiO_2$) and low Ti (<2.5 wt.% $TiO_2$) magnetite. High-Ti magnetite is usually accompanied by significantly to relatively depleted $Fe_2O_3$ (30-60 wt.%) contents relative to low-Ti magnetite (>55 wt.% $Fe_2O_3$) (Figs. 5.1A and B). High-Ti magnetite is found solely within massive mineralized samples (i.e. massive oxides and nelsonites) from Buttercup (7.9-18.4 wt.% $TiO_2$), St. Charles de Bourget (2.6-14.6 wt.% $TiO_2$) and Lac Perron (1.4-6.6 wt.% $TiO_2$). However, an exception to this rule exists for massive oxide samples from Lac Margane/Houlière-Nord and Lac à l'Orignal which contain unusually low-Ti magnetite (< 0.5 wt.% $TiO_2$). Within all of the disseminated oxide samples only low-Ti magnetite is present (Figs. 5.1A and B): Lac Perron (1.4 wt.% $TiO_2$), Lac Margane (0.9-0.1 wt.% $TiO_2$), New Zone of Lac à Paul (0.2-2.5 wt.% $TiO_2$), and Lac à l'Orignal (<0.1 wt. % $TiO_2$) deposits and Buttercup host anorthosite (0.6 wt.% $TiO_2$).

Similar to magnetite, ilmenite can also be separated into groups based on average $Fe_2O_3$ content (Fig. 5.2): 1) $Fe_2O_3$-poor ilmenite ($Fe_2O_3$ < 3 wt.%), 2) $Fe_2O_3$-enriched ilmenite ($Fe_2O_3$ 3-10 wt.%), and 3) hemo-ilmenite ($Fe_2O_3$ > 10 wt.%). These groupings were created in combination with petrography since some ilmenite can have elevated $Fe_2O_3$ contents but not contain hematite exsolutions. All analyses of ilmenite with obvious hematite exsolutions (Fig. 5.2) are hemo-ilmenite (only contained $Fe_2O_3$ contents >10 wt.%) and occur only in the disseminated nelsonitic norite at Lac à l'Orignal (15.8-17.4 wt.% $Fe_2O_3$). $Fe_2O_3$-poor ilmenite (< 3 wt.% $Fe_2O_3$) is the dominant species within the massive oxide samples, except for those in the massive samples from Lac Margane/Houlière-Nord, Lac à Paul ore zone (data from

Fredette 2006) and Lac à l'Orignal which are enriched in $Fe_2O_3$ (8 wt.% $Fe_2O_3$). All ilmenite

grains analysed within disseminated oxide samples from the Lac Margane/Houlière-Nord

deposit also belong within the $Fe_2O_3$-enriched group (5.2-9.3 wt.% $Fe_2O_3$). However, ilmenite

from samples containing semi-massive/disseminated oxides from both Lac Perron (MG-LPR-

08; 1.1 wt% $Fe_2O_3$) and the New Zone of Lac à Paul (1.2-3.9 wt. %; Fredette 2006) are mostly

$Fe_2O_3$-poor. There also appears to be a geographical control on the $Fe_2O_3$ content of ilmenite:

mineralization from the South and West (Buttercup, St. Charles de Bourget and Lac Perron)

contain $Fe_2O_3$-poor ilmenite whereas mineralization from the North and East (Lac

Margane/Houlière-Nord, Lac à Paul, and Lac à l'Orignal) contain $Fe_2O_3$-enriched ilmenite.

Analyses on Al-spinel could only be performed on samples containing massive/semi-

massive oxides as it was rare/absent in samples containing disseminated oxides. The

composition of Al-spinel displays significantly less variation than the composition of

magnetite and ilmenite. All Al-spinel grains analyzed are pleonaste (Fe-Mg-Al spinel) in

composition, which is intermediate between hercynite ($FeAl_2O_4$ spinel) and spinel ($MgAl_2O_4$

spinel) end members of the spinel supergroup. However, FeO and MgO contents can vary

among the Al-spinels from each deposit/occurrence. The Al-spinels at Lac

Margane/Houlière-Nord are most depleted in MgO (6.6-7.7 wt.% MgO) and enriched in FeO

(27.8-29.6 wt.%) whereas Al-spinel at Buttercup is the most enriched in MgO (8.8-14.5 wt.%

MgO) and depleted in FeO (19.20-27.03 wt.% FeO). Concentrations of other elements within

spinel remained relatively constant except for within one sample at Lac Margane/Houlière-

Nord which contained anomalously high $Cr_2O_3$ contents (3.3 wt.%).

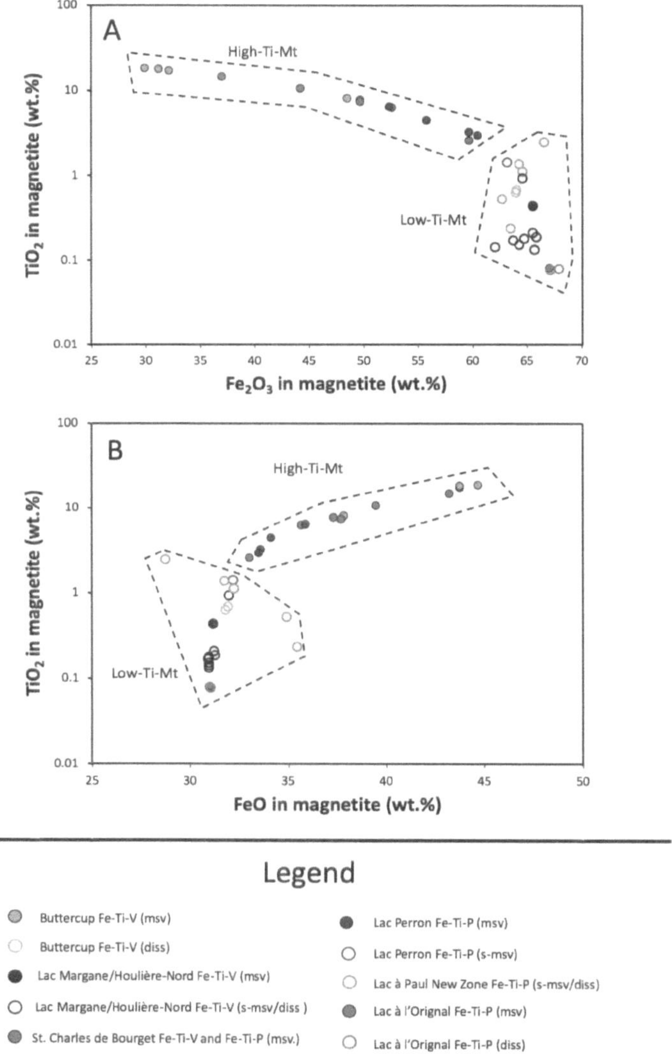



**Figure 5.1:** Binary plots displaying the average $TiO_2$ content and Fe speciation obtained by EMPA within magnetite from each location of Fe-oxide ± apatite mineralization examined in this study (Fe-oxide ± apatite content noted in brackets). A) $TiO_2$ vs $Fe_2O_3$. B) $TiO_2$ vs. FeO content. Msv=massive, s-msv=semi-massive, diss=disseminated.

83

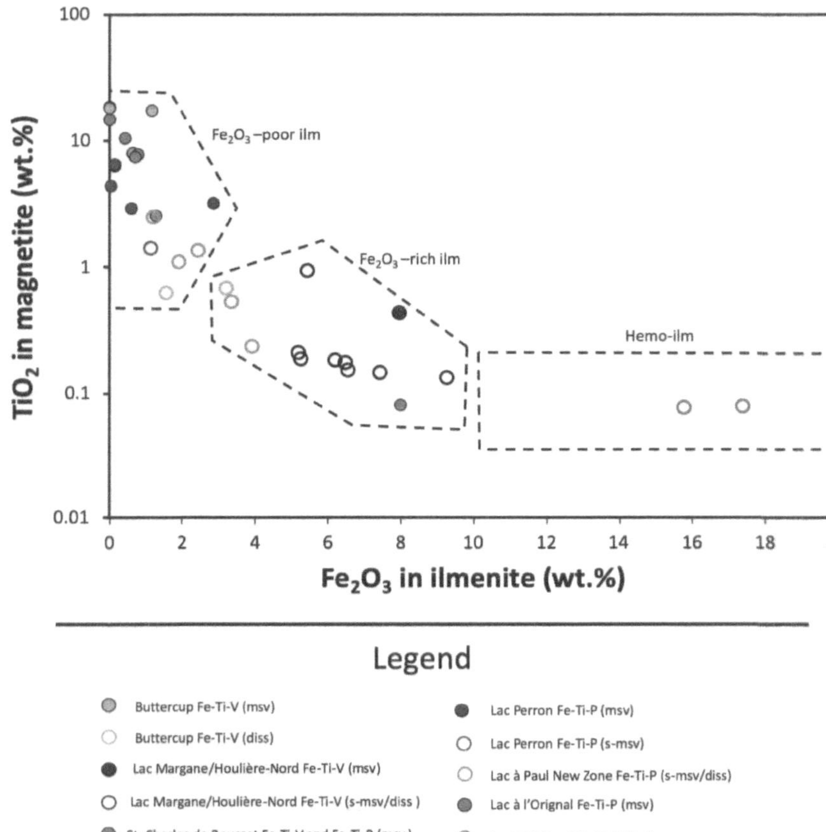


## Legend

| | |
|---|---|
| ◉ Buttercup Fe-Ti-V (msv) | ● Lac Perron Fe-Ti-P (msv) |
| ○ Buttercup Fe-Ti-V (diss) | ○ Lac Perron Fe-Ti-P (s-msv) |
| ● Lac Margane/Houlière-Nord Fe-Ti-V (msv) | ○ Lac à Paul New Zone Fe-Ti-P (s-msv/diss) |
| ○ Lac Margane/Houlière-Nord Fe-Ti-V (s-msv/diss ) | ◉ Lac à l'Orignal Fe-Ti-P (msv) |
| ◉ St. Charles de Bourget Fe-Ti-V and Fe-Ti-P (msv.) | ○ Lac à l'Orignal Fe-Ti-P (diss) |

**Figure 5.2:** Binary plot displaying average TiO₂ in magnetite vs. Fe₂O₃ contents of ilmenite obtained by EMPA from each location of Fe-oxide ± apatite mineralization examined in this study (Fe-oxide ± apatite content noted in brackets). Msv=massive, s-msv=semi-massive, diss=disseminated.

## 5.3 Major and minor element composition of silicate minerals

The primary igneous silicates that were analyzed during this project are plagioclase, orthopyroxene, and olivine. The results shown in this section (displayed on Table 5.1) are interpreted as partially representative of cumulus compositions due to possible effects of re-equilibration and metamorphism, except for olivine symplectites at Buttercup (Fig. 3.5A) and orthopyroxene reaction coronas around olivine at Lac à Paul (New Zone: Fig. 4.13D), which were also analyzed. Secondary silicate minerals, such as amphibole, biotite, and garnet formed by subsolidus reactions between magnetite and primary silicate minerals (plagioclase, olivine, orthopyroxene) related to sub-solidus processes or metamorphism were also analysed.

### 5.3.1 Primary silicates

In general, plagioclase is the most common silicate mineral in disseminated oxide samples and host mafic rocks. The Anorthite content (An) of plagioclase, using the molar ratio $\frac{Ca}{(Ca+Na+k)}$, is typically used as a robust indicator of evolution of the magma in mafic intrusions, where An content decreases during differentiation. In general, plagioclase present in disseminated oxide samples and the host mafic rocks of the 2 Fe-Ti-V deposits/occurrences studied are mainly of labradorite-type ($An_{48-60}$) while those from the 4 Fe-Ti-P deposits/occurrences are mainly of andesine-type and locally (at Lac à Paul-New Zone) oligoclase type ($An_{25-47}$) (Fig. 5.2). This is similar to what was observed in a regional study of magmatic oxide mineralization (Fe-Ti-V-P) within the Lac St. Jean anorthosite suite by Hébert et al. (2005; 2009). Anomalously low An values ($An_{43-45}$) at Lac Margane/Houlière-Nord is likely due to the presence of garnet within these samples.

**Table 5.1:** Habits and compositions of silicate minerals (plagioclase, olivine and orthopyroxene) within massive to disseminated oxide samples and host rocks examined in this study. Relict (orthopyroxene) implies primary crystal which has been partially converted to amphibole. An = anorthite, Pl = plagioclase, Ol = olivine, Fo = Forsterite, Opx= orthopyroxene, Msv. = massive, S-msv. = semi-massive, 1ry = primary, w = with, nph = nepheline.

| Deposit | Sample | Lithology | Pl texture | An Content (Pl) | Ol texture | Fo Content (Ol) | Opx texture | Mg# (Opx) |
|---|---|---|---|---|---|---|---|---|
| Buttercup (Fe-Ti-V) | MG-BCP-1a | Anorthosite | 1ry crystal | 49 | - | - | - | - |
| | BCP-2 | Msv. Oxide | - | - | symplectite w nph | 71 | - | - |
| | MG-BCP-8 | Anorthosite | 1ry crystal | 50 | - | - | - | - |
| | MG-BCP-9 | Anorthosite | 1ry crystal | 51 | - | - | - | - |
| Lac Margane/Houlière-Nord (Fe-Ti-V) | MA-01c | Fe-oxide norite | 1ry crystal | 58 | - | - | - | - |
| | MA-02b | Fe-oxide norite | 1ry crystal | 60 | - | - | - | - |
| | WP-207 | Fe-oxide norite | 1ry crystal | 45 | - | - | - | - |
| St. Charles de Bourget (Fe-Ti-P and Fe-Ti-V) | MG-SCB-1b | Msv. Oxide | - | - | xenolith | 57 | - | - |
| | MG-SCB-3 | Msv. Oxide | xenolith | 43 | - | - | - | - |
| | MG-SCB-12 | Msv. Oxide | xenolith | 44 | xenolith | 43 | - | - |
| | MG-SCB-14 | Nelsonite | - | - | xenolith | 59 | - | - |
| Lac Perron (Fe-Ti-P) | MG-LPR-7 | Anorthosite | 1ry crystal | 43 | - | - | - | - |
| | MG-LPR-8 | S-msv. Nelsonite | 1ry crystal | 35 | - | - | - | - |
| Lac à Paul New Zone (Fe-Ti-P) | SDLP-18 | dunitic nelsonite | - | - | 1ry crystal | 59 | Corona on Ol | 67 |
| | SDLP-36a | nelsonitic olivine norite | 1ry crystal | 34 | 1ry crystal | 53 | 1ry crystal | 62 |
| | SDLP-36b | nelsonitic olivine norite | 1ry crystal | 41 | - | - | 1ry crystal | 52 |
| | SDLP-37 | nelsonitic pyroxenite | - | - | - | - | relict | 67 |
| | SDLP-38 | nelsonitic norite | 1ry crystal | 26 | - | - | relict | 41 |
| Lac à l'Orignal (Fe-Ti-P) | LacOrignal | nelsonitic norite | 1ry crystal | 47 | - | - | 1ry crystal | 63 |
| | MG-LO-1 | nelsonitic norite | 1ry crystal | 44 | - | - | 1ry crystal | 65 |
| | MG-LO-2b | anorthosite/leuconorite | 1ry crystal | 44 | - | - | 1ry crystal | 62 |
| | MG-LO-3 | nelsonitic norite | 1ry crystal | 41 | - | - | 1ry crystal | 65 |

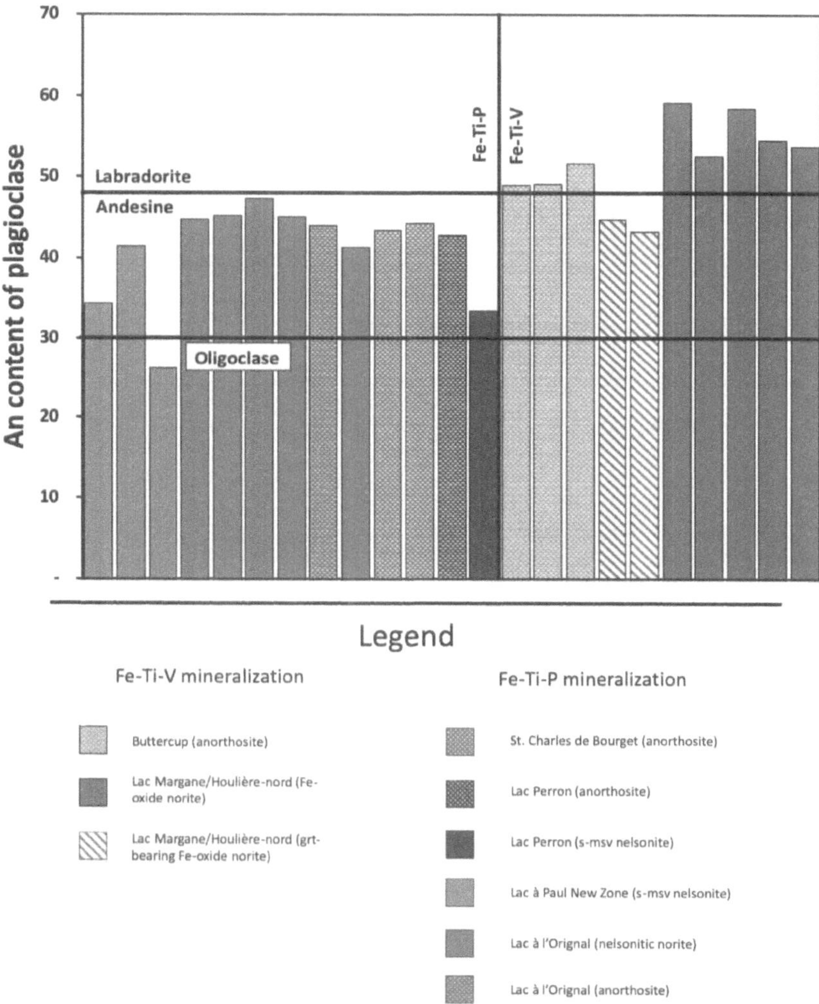



**Figure 5.3:** Bar graph displaying the Anorthite (An) content of plagioclase obtained by EMPA from within samples of host rocks and disseminated oxide samples from each deposit/occurrence studied in this project. Grt=garnet, S-msv=semi-massive.

Orthopyroxene was analyzed by EMPA from the Lac à Paul (New Zone) and Lac à l'Orignal mineralization (Table 5.1). From Lac à Paul (New Zone), primary orthopyroxene was analyzed from a nelsonitic olivine norite, nelsonitic pyroxenite, and a nelsonitic norite while secondary orthopyroxene, which forms coronas around olivine, was analyzed within a dunitic nelsonite. From Lac à l'Orignal, primary orthopyroxene was analyzed within the mineralized nelsonitic norite and an anorthosite/leuconorite host rock. The Mg#, using the molar ratio $(\frac{Mg}{Mg+Fe})$ , of pyroxene is typically used as a robust indicator of magmatic evolution within mafic intrusions as the Mg# should decrease during differentiation. However, Mg# within Fe-Mg silicates (i.e. orthopyroxene and olivine) can also increase due to sub-solidus re-equilibration of Mg and Fe with Fe-oxides as shown by previous work (Morse 1980; Pang et al. 2007; Charlier et al. 2015).

At Lac à Paul (New Zone), orthopyroxene with the highest Mg# is found within the nelsonitic pyroxenite unit (Mg# = 67) whereas orthopyroxene with the lowest Mg# occurs within the nelsonitic norite unit (Mg# = 41). The nelsonitic olivine-norite, which was separated into two samples based on oxide content (SDLP-36a: oxide >20% and SDLP-36b: oxide <20%), contains orthopyroxene with Mg# of 62 and 52, respectively, illustrates the increase in Mg# with increasing content of Fe-oxides. However, the difference in Mg# between the nelsonitic pyroxenite and nelsonitic norite units, which contain similar contents of Fe-oxides (<20%), can potentially be related to magmatic differentiation. Orthopyroxene present as reaction coronas surrounding olivine within the dunitic nelsonite (SDLP-18) has Mg#=67. However, since these grains formed during secondary processes, the results are not indicative of magmatic evolution. Analyses of primary orthopyroxene from Lac à l'Orignal show little significant change in its composition as Mg# ranges from 62-65. However, orthopyroxene from the nelsonitic norites from Lac à l'Orignal have elevated Mg# relative to nelsonitic norites from Lac à Paul, perhaps indicating a more primitive composition. This is also shown by elevated An contents within plagioclase from nelsonitic norites at Lac à l'Orignal ($An_{41-47}$) relative to Lac à Paul-New Zone ($An_{26-41}$).

Primary olivine is present within dunitic nelsonites and olivine gabbronorites from Lac à Paul (New Zone) and as xenoliths of dunite within both massive oxide and massive nelsonitic mineralization from St. Charles de Bourget. The Fo (forsterite) content, or the Mg#, using the molar ratio ($\frac{Mg}{Mg+Fe}$), of olivine is generally used as a robust indicator of magmatic evolution within olivine-bearing igneous rocks and Fo should decrease during differentiation. However, the Fo content of olivine can also increase during sub-solidus re-equilibration with Fe-oxides similarly to Mg# in orthopyroxene. Therefore, sub-solidus re-equilibration must be considered in the final interpretation of olivine compositions since most samples containing olivine contain >30% Fe-oxides. Olivine from the dunitic nelsonite (oxides >40%) and olivine gabbro-norite oxides <20%) at Lac à Paul (New Zone) have compositions of $Fo_{57}$ and $Fo_{53}$ respectively. Olivine from massive oxides at St. Charles de Bourget range from $Fo_{57}$ in the first sampling area to $Fo_{43}$ at the second sampling area (cottage outcrop; Fig. 4.2). Dunite xenoliths from the nelsonite outcrop contain olivine with the highest Fo contents ($Fo_{59}$), which is a similar composition to olivine from the first sampling area. Symplectic intergrowth of secondary olivine with nepheline from within the Buttercup massive ore has a composition of $Fo_{71}$, which is abnormally high for Fe-oxide mineralization, and is interpreted to be attributed to re-equilibration processes.

5.3.2   Silicate minerals generated from reactions with Fe-oxides

Additional secondary silicate minerals found among the deposits/occurrences examined in this study are amphibole, biotite, and garnet (Appendix 4). These minerals are generally found within semi-massive or disseminated lithologies containing both plagioclase and Fe-oxides, such as Fe-oxide norites (Lac Margane/Houlière-Nord), nelsonitic norites (Lac à Paul-New Zone and Lac à l'Orignal), anorthosites with trace oxides (Buttercup). Based on the classifications by Hawthorne et al. (2012), most of the amphiboles (reaction coronas and granular amphibole) measured by EMPA are of the hornblende variety (specifically pargasite) with 2 exceptions; 1) cummingtonite was observed as the tabular amphibole crystals co-existing with blocky hornblende in Fe-oxide norites at Lac Margane/Houlière-Nord (Fig 3.11D) and 2) Ti-pargasite was observed as the brownish amphibole crystals which formed as coronas surrounding the symplectic olivine and nepheline and as crystals in contact between plagioclase and Fe-oxides at Buttercup (Fig 3.5A).

Biotite analysed during this study can be grouped into two types: 1) high-Ti (>2.5 $TiO_2$ wt.%) and low-Ti (<2 $TiO_2$ wt.%), with high-Ti biotite being the most common. High-Ti biotite is usually noted within coronas surrounding Fe-oxide minerals and low-Ti biotite is usually distal from the Fe-oxides and more closely associated with granular amphiboles. All garnets, analyzed by EMPA, in the disseminated oxides from Lac Margane/Houlière-nord anorthosite near mineralization contact from Buttercup, and anorthosite near mineralization from Lac Perron have almandine-dominant compositions (Fe-rich).

5.4   Volatile content of Apatite

While the EMPA analyses show that all the apatites analyzed from the 4 Fe-Ti-P mineralization are fluoro-apatites, the volatile content does show some variation among each deposit/occurrence. To volatile contents are compared according to rock type (massive vs.

disseminated nelsonitic samples; Fig. 5.4). Apatite from nelsonites refers to St. Charles de Bourget, Lac Perron and Lac à Paul ore zone (data from Desormiers, 2015) whereas apatite from nelsonitic norites refers to Lac à l'Orignal and Lac à Paul (New Zone). Apatite in nelsonitic rocks from the 3 different Fe-Ti-P deposits/occurrences in the 1.15 Ga Lac St. Jean anorthosite suite generally have similar volatile contents (2.8-3.3 wt.% F and 0.08-0.1 wt.% Cl; Fig. 5.4A) and F/Cl ratios between 24.1 –88 % (Fig 5.4B). However, disseminated apatite from the Lac à Paul New Zone (apatite-poor, nelsonitic norites and pyroxenites) can have elevated Cl contents (< 0.08 wt.%), and in the case of the pyroxenite sample, lower F contents (1.9-2.3 wt.%) which causes depleted F/Cl ratios (13.5-60.5). In contrast apatite from the apatite-rich nelsonitic norites from the Lac à l'Orignal Fe-Ti-P deposit hosted within the 1080 ± 2 (van Breeman et al. 2009) Lac Vanel anorthosite suite is different from apatite from the older Lac St Jean mineralization. It is more generally enriched in both F (3.2-3.8 wt.%; Fig. 5.4A and B) and Cl (0.08-0.13 wt.%; Fig. 5.4A).

Volatile anion contents of apatite are plotted on the ternary diagram F-Cl-OH in Fig. 5.5 which shows significant variation within OH contents of apatite among the deposits/occurrences. Apatite from younger Lac à l'Orignal nelsonitic norites contain the most depleted OH contents whereas apatite from Lac à Paul (Paul Zone and New Zone) contains the most enriched OH contents with apatite from the pyroxenite sample (New Zone) being the most enriched (Fig. 5.5). Within the massive nelsonite ores, apatite from Lac Perron is depleted in OH relative to apatite from St. Charles de Bourget (Fig. 5.5). Overall, the high F, negligible Cl and low OH contents of all the mineralised samples are similar to the apatite field from the Skaergaard and Munni Munni intrusions rather than the high-Cl apatites of the Bushveld and Stillwater lower zones (which are intercumulus apatite).

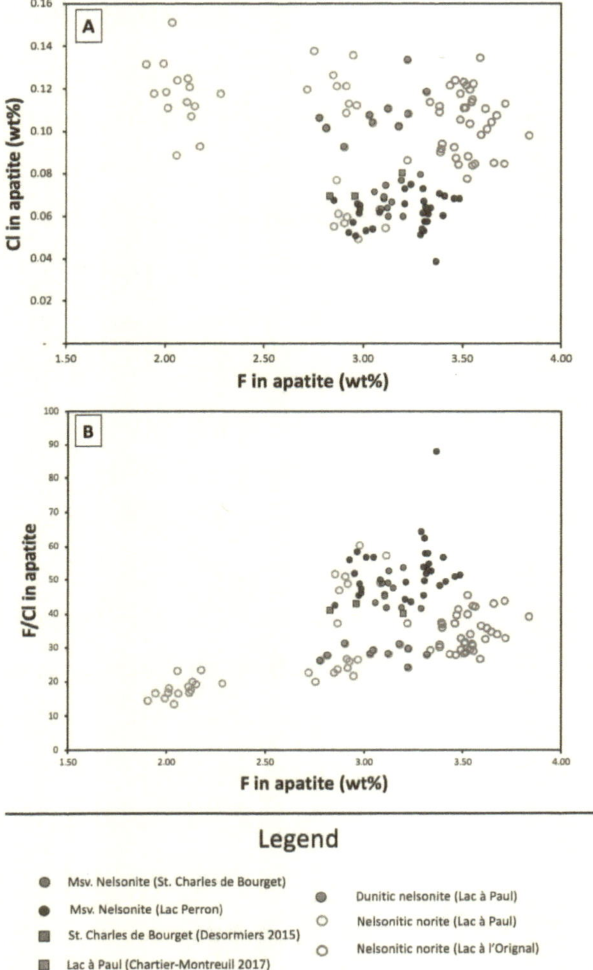



**Figure 5.4:** Binary plots displaying volatile contents of fluoro-apatite obtained by EMPA. A) Cl vs. F content in apatite. B) F/Cl vs. F in apatite. Circles = this study (individual analyses); Squares = averages of apatite analyses from previous work: St. Charles de Bourget (Desormiers, 2015) and Lac à Paul (Chartier-Montreuil 2017).

92

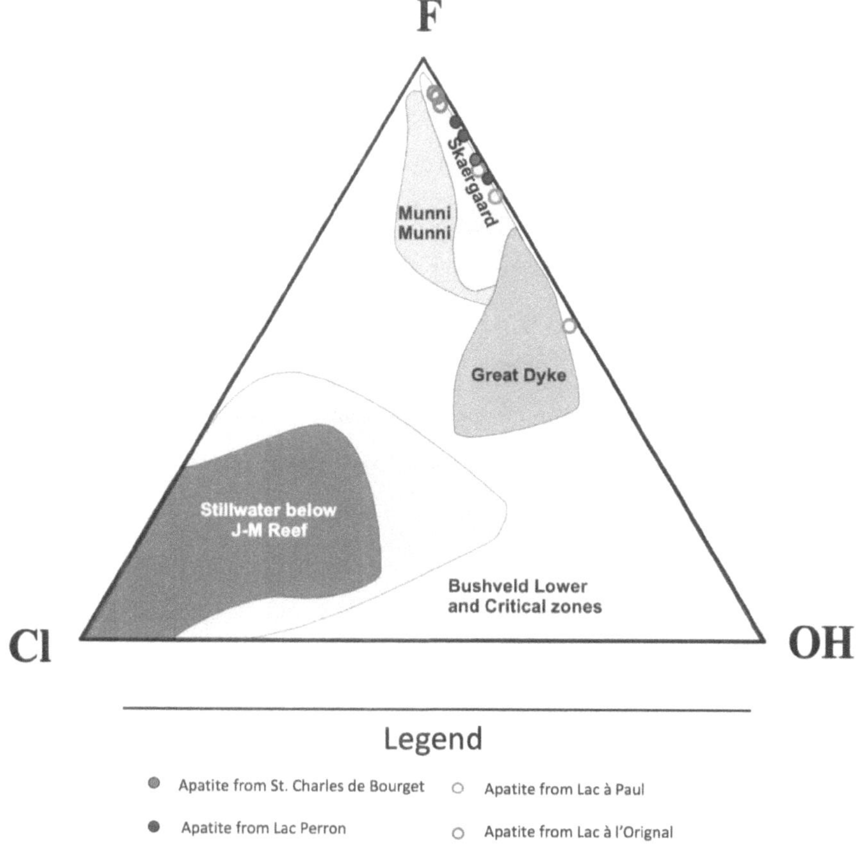


**Figure 5.5:** Ternary diagram displaying the volatile content (F-Cl-OH) of apatite within mineralization from this study. Results are plotted against data fields that display the volatile content of apatite from several well-known mafic-ultramafic intrusions associated with large igneous provinces. Taken from Boudreau et al. (1995).

93

# Chapter 6   Trace element compositions of Fe-oxides

This section includes the analytical protocol of in-situ LA-ICP-MS analyses on magnetite, ilmenite, and Al-spinel comparison of LA-ICP-MS data with EMPA data, partitioning behavior of elements between the Fe-oxide phases, and the evaluation of trace element trends within magnetite. Fe-oxide grains were analyzed from the 2 Fe-Ti-V deposits/occurrences (Buttercup and Lac Margane/Houlière-Nord) and the 4 Fe-Ti-P deposits/occurrences (1) St. Charles de Bourget, 2) Lac Perron, 3) Lac à Paul, and 4) Lac à l'Orignal. The Fe-oxide results displayed in this section are interpreted based on the sampling location (mineralization from which the Fe-oxide phase was sampled and location within the mineralization) and oxide content (i.e. massive, semi-massive, or disseminated Fe-oxides ± apatite), since Fe-oxides are present within all lithologies examined in this study. Overall, the results presented in this section aim to provide new constraints on the composition of Fe-oxides within the different mineralization styles and host rocks in the Lac St. Jean area.

## 6.1   Analytical Methods

The Fe-oxide phases, analyzed in 32 samples from the 6 deposits/occurrences listed above by LA-ICP-MS, include magnetite (n=157 grains), ilmenite (n=170), and Al-spinel (n=63). LA-ICP-MS analysis was performed at the University of Ottawa using a Photon Machines Analyte (193nm) Excimer laser system and a 7700 Agilent quadropole ICP-MS and analytical conditions are summarized in Tables 6.1 and 6.2. Ablation methods for Fe-oxides involved using a beam diameter of 52μm (Fig. 6.1) to ablate 300 μm long lines during a 60s acquisition time (at a scan speed of 5 μm/s). Before and after acquisition, a 30s gas blank was analyzed to constrain the background. Approximately 3-6 grains of each Fe-oxide phase were analyzed, if present in high enough modal percentages. Magnetite was regularly present in each sample to perform 3-6 analyses but, ilmenite and Al-spinel modal percentages were only high enough to perform a single analysis in some samples.

The element lists were the same for each Fe-oxide phase and consisted of 25 trace elements (Mg, Al, Si, P, Ca, Sc, Ti, V, Cr, Mn, Co, Ni, Cu, Zn, Ga, Ge, Y, Zr, Nb, Mo, Sn, Hf, Ta, W, and Pb) following the protocol of Dare et al. (2014). Calibration was performed using certified USGS reference material GSE-1g, which is a synthetic glass with approximately 10% Fe and approximately 500 ppm of each listed trace element (Jochum et al. 2005). Analysis of up to 10-12 Fe-oxide grains (unknowns) were analysed between repeated analyses of GSE-1g followed by 4 other reference materials to monitor data quality (Table 6.3): GSD-1g (Fe-rich synthetic glass doped with 40-50 ppm of each trace element, Jochum et al. 2005), BCR-2g (natural basaltic glass, Jochum et al. 2006), GOR-128g (natural komatiitic glass, Jochum et al 2006), and in-house standard BC-28 (magnetite from the main magnetite layer of the Bushveld Complex, Dare et al. 2014). Data reduction was performed by Glitter® using $^{57}$Fe as the internal standard for Fe- oxides, determined by electron microprobe, while Si, P, S, Ca, and Cu were used to monitor inclusions during signal processing.

Analyses of the certified reference materials are presented in Table 6.3. Results from GSD-1g are accurate and precise; all elements have relative differences (RD) below 10% and have relative standard deviations (RSD) below 5%. With BCR-2g, elements are relatively accurate (<10% RD) with the exceptions of Cu, Zn, Ga, Ge, Y, and Ta; there are several elements with relative standard deviations above 10% which include those mentioned and Sc, Cr, Ni, Nb, Sn, W, and Pb. All elements mentioned above, except Zn, have concentrations <100ppm within BCR-2g. Analyses of GOR-128g produce accurate and precise results for all elements with mass numbers below $^{71}$Ga. This is because heavier elements are generally present in very low concentrations (<1ppm - 16ppm) in this reference material. Generally, analyses of reference materials are accurate and precise if elemental concentrations are relatively one orders of magnitude larger than their detection limits.

Analyses of BC-28 are given in Table 6.4. Since BC-28 is a natural massive magnetite sample, containing fine-grained exsolutions of Al-spinel, it is not expected to be as homogeneous as glass reference materials but it gives a good indication if the calibration method works for natural magnetite. As such, relative difference (accuracy) of 15 -20% is considered acceptable (Dare et al. 2014). Repeated LA-ICP-MS analysis of BC-28 produces measurements that are precise (< 10% RSD) and relatively accurate for most elements in concentration above 30 ppm, except for Si, Co, Cu, and Zn. This is likely due to heterogeneous distribution of these elements caused by incorporation of micro inclusions or exsolutions into the analyses that were not obvious during data reduction. Analyses of BC-28 by EMPA agree with analyses by LA-ICP-MS and working values for Ti, Cr, Mn, and Ni (Table 6.4). Concentrations of V within BC-28 obtained by EMPA consistently display lower concentrations than the working values perhaps due to the correction for the interference of Ti on V by EMPA. Therefore, the LA-ICP-MS data is preferred over the EPMA data for V. In contrast, analyses of Al within BC-28 by EMPA consistently display higher concentrations relative to the working values. This is likely because the small analyses area (20 μm spot) by EMPA can possibly lead to increased signal from Al-spinel exsolutions compared to LA-ICP-MS (approximately 300 μm X 50 μm line). Varying signal of Al-spinel exsolutions during EMPA analyses of BC-28 could also be responsible for analyses of Mg and Zn which are neither accurate nor precise.

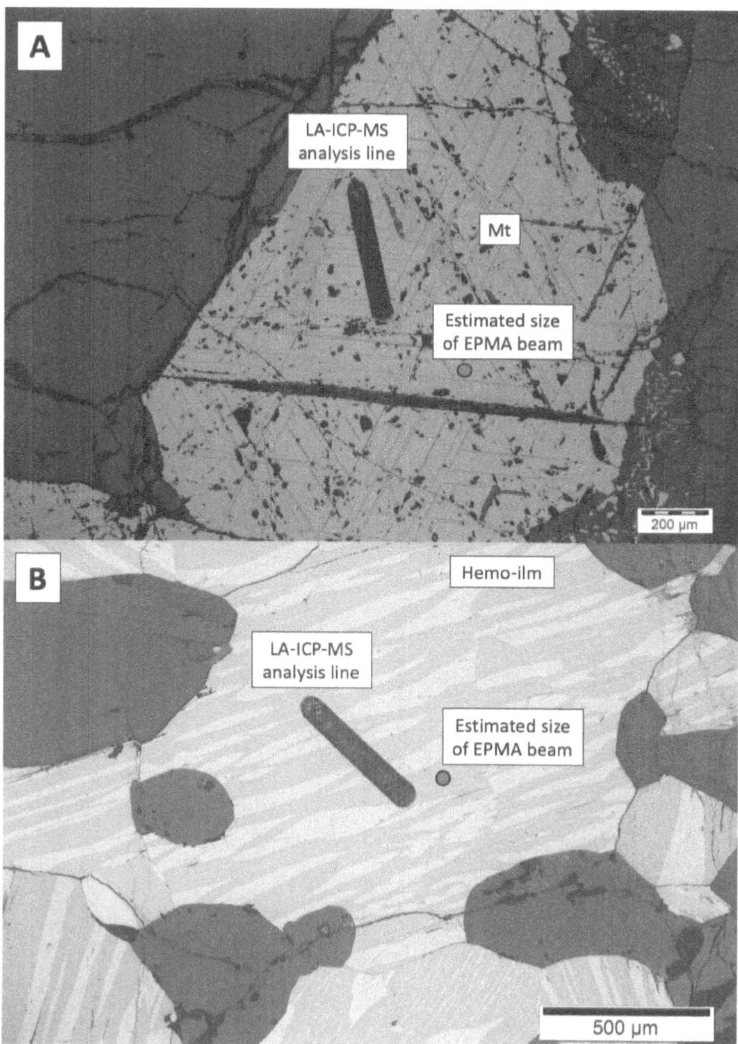

**Figure 6.1**: Photomicrographs of exsolution-rich Fe-oxide grains displaying the relative analysis area covered by LA-ICP-MS (50 µm diameter, ~300 µm length line) compared to EMPA (20 µm spot). Note the much larger area of analysis on the crystal by LA-ICP-MS and that the LA-ICP-MS ablation line is oriented almost perpendicular to coarse include exsolution phases while the EMPA beam avoids coarse exsolutions. A) Magnetite (mt) grain containing exsolutions of Al-spinel. B) Hemo-ilmenite (Hemo-ilm) grain from Lac à l'Orignal.

**Table 6.1**: Analytical details for LA-ICP-MS analysis of Fe-oxides.

| | University of Ottawa |
|---|---|
| Laser Ablation System | Photon-Machines Analyte 193nm Excimer |
| ICP-MS | Agilent 7700x |
| Laser Frequency | 15 Hz |
| Pulse Energy | 5 mJ/pulse |
| Stage Speed | 5 µm/s |
| Beam Diameter | 52 µm |
| Dwell Time | See Table 6.2 |
| Analysis Procedure | 30 s of gas blank 60 s of signal |
| Carrier Gas Flow | Helium (1/L/min) |
| Ar Flow Rate | 0.7 L/Min |
| Internal Standard | Fe (data from EPMA) |
| Reference Material for Calibration | GSE-1g (USGS-synthetic glass) |
| Reference Material for Monitoring | GSD-1g (USGS-synthetic glass) |
| | BCR-2g (Basaltic natural glass) |
| | GOR-128g (Komatiitic natural glass) |
| In-house Monitor | BC-28 (Magnetite from the Bushveld Complex) |
| Monitor Signal for Inclusions | Si, S, P, Ca, Cu |

**Table 6.2**: Isotopic dwell times used during LA-ICP-MS analysis at the University of Ottawa.

| Isotope | | Dwell Time (ms) |
|---|---|---|
| 24 | Mg | 5 |
| 27 | Al | 5 |
| 29 | Si | 5 |
| 31 | P | 5 |
| 34 | S | 5 |
| 43 | Ca | 5 |
| 45 | Sc | 10 |
| 47 | Ti | 5 |
| 51 | V | 10 |
| 52 | Cr | 10 |
| 55 | Mn | 10 |
| 57 | Fe | 10 |
| 59 | Co | 10 |
| 60 | Ni | 10 |
| 63 | Cu | 20 |
| 66 | Zn | 10 |
| 69 | Ga | 10 |
| 74 | Ge | 20 |
| 89 | Y | 20 |
| 90 | Zr | 10 |
| 93 | Nb | 10 |
| 95 | Mo | 20 |
| 118 | Sn | 15 |
| 178 | Hf | 10 |
| 181 | Ta | 10 |
| 182 | W | 10 |
| 208 | Pb | 10 |

**Table 6.3**:Results of LA-ICP-MS analyses of reference materials. All element concentrations are given in ppm

| | Detection Limits | | Calibration GSE-1g | | Monitor GSD-1g | | | | Monitor BCR-2G | | | | Monitor GOR-128g | | | |
| | LA-ICP-MS | EPMA | Certificate Values | | Certificate Values | | uOttawa (this study) | | Certificate Values | | uOttawa (this study) | | Certificate Values | | uOttawa (this study) | |
| | | | | | | | Average | | | | Average | | | | Average | |
| Isotope | 52µm | 20µm | Average | St.Dev | Average | St.Dev | (n=14) | St.Dev | Average | St.Dev | (n=20) | St.Dev | Average | St.Dev | (n=28) | St.Dev |
|---|---|---|---|---|---|---|---|---|---|---|---|---|---|---|---|---|
| Mg24 | 0.1 | 89 | 21106 | 181 | 21709 | 241 | 21817 | 170 | 21654 | 603 | 20790 | 269 | 156826 | 1810 | 151017 | 1763 |
| Al27 | 0.2 | 62 | 68804 | 2117 | 70922 | 1588 | 71499 | 693 | 71446 | 2117 | 68740 | 1476 | 52446 | 900 | 49299 | 943 |
| Si29 | 43 | 55 | 250994 | 7011 | 248657 | 3739 | 254245 | 5014 | 252959 | 7481 | 249579 | 15651 | 215553 | 1870 | 207770 | 12619 |
| Ca44 | 16 | 65 | 52858 | 2143 | 51429 | 714 | 52702 | 775 | 50856 | 1571 | 50941 | 771 | 44571 | 857 | 43640 | 685 |
| Sc45 | 0.03 | - | 530 | 20 | 52 | 2 | 56 | 3 | 33 | 4 | 35 | 1 | 32 | 1 | 31 | 1 |
| Ti47 | 0.07 | 198 | 450 | 42 | 7432 | 360 | 8059 | 177 | 13548 | 599 | 14115 | 515 | 1727 | 72 | 1653 | 44 |
| V51 | 0.01 | 90 | 440 | 20 | 44 | 2 | 45 | 0.5 | 416 | 28 | 447 | 8 | 189 | 13 | 190 | 3 |
| Cr52 | 0.1 | 170 | 400 | 80 | 42 | 3 | 46 | 0.8 | 18 | 4 | 17 | 0.2 | 2272 | 171 | 2338 | 41 |
| Mn55 | 0.06 | 106 | 590 | 20 | 220 | 20 | 223 | 3 | 1520 | 120 | 1545 | 17 | 1363 | 70 | 1359 | 14 |
| Co59 | 0.01 | - | 380 | 20 | 40 | 2 | 40 | 0.4 | 37 | 6 | 37 | 0.3 | 92 | 6 | 86 | 0.7 |
| Ni60 | 0.8 | 116 | 440 | 30 | 58 | 4 | 58 | 2 | 13 | 4 | 12 | 0.2 | 1074 | 61 | 1040 | 14 |
| Cu63 | 0.03 | - | 380 | 40 | 42 | 2 | 42 | 0.7 | 21 | 10 | 17 | 0.4 | 64 | 13 | 61 | 1 |
| Zn66 | 0.1 | 525 | 460 | 10 | 54 | 2 | 54 | 1 | 127 | 18 | 146 | 2 | 75 | 7 | 72 | 2 |
| Ga69 | 0.02 | - | 490 | 70 | 54 | 7 | 53 | 1 | 23 | 4 | 32 | 0.9 | 9 | 1 | 8 | 0.2 |
| Ge74 | 0.05 | - | 320 | 80 | 32 | 8 | 32 | 0.5 | 1.5 | 0.2 | 1.3 | 0.0 | 1 | 0.05 | 1 | 0.05 |
| Y89 | 0.001 | - | 410 | 30 | 42 | 2 | 42 | 1 | 37 | 4 | 33 | 1 | 12 | 0.5 | 11 | 0.5 |
| Zr90 | 0.01 | - | 410 | 30 | 42 | 2 | 44 | 1 | 188 | 32 | 184 | 7 | 10 | 0.5 | 10 | 0.6 |
| Nb93 | 0.002 | - | 420 | 40 | 42 | 3 | 42 | 1 | 13 | 2 | 11 | 0.3 | 0.1 | 0.01 | 0.09 | 0.02 |
| Mo95 | 0.01 | - | 390 | 30 | 39 | 3 | 40 | 0.6 | 248 | 34 | 264 | 3 | 0.7 | 0.03 | 0.6 | 0.05 |
| Sn118 | 0.3 | - | 280 | 50 | 29 | 6 | 31 | 0.6 | 3 | 0.8 | 3 | 0.3 | 0.2 | 0.09 | 2 | 0.9 |
| Hf178 | 0.003 | - | 395 | 7 | 39 | 2 | 41 | 1 | 5 | 0.6 | 5 | 0.2 | 0.4 | 0.02 | 0.3 | 0.03 |
| Ta181 | 0.002 | - | 390 | 40 | 40 | 4 | 40 | 1 | 0.8 | 0.1 | 0.7 | 0.0 | 0.02 | 0.00 | 0.02 | 0.005 |
| W182 | 0.006 | - | 430 | 50 | 43 | 4 | 44 | 0.5 | 0.5 | 0.1 | 0.5 | 0.0 | 16 | 2 | 16 | 0.5 |
| Pb208 | 0.01 | - | 378 | 12 | 50 | 2 | 50 | 1 | 11 | 2 | 11 | 0.2 | 0.4 | 0.04 | 0.4 | 0.02 |

[57]Fe was used as the internal standard for all analyses using the following values: 9.8 wt.% (GSE-1g), 10.3 wt.% (GSD-1g), 9.6 wt.% (BCR-2g), and 7.6 wt.% (GOR-128g). Certificate and working values taken from Dare et al. (2014).

**Table 6.4:** Results of LA-ICP-MS and EMPA analyses of in-house reference material BC-28 (natural magnetite). All element concentrations are given in ppm. BC-28 contains fine-grained Al-spinel trellis exsolutions which cause increased relative differences and relative standard deviation by EMPA.

| | Detection Limits | | In-house Monitor BC-28 LA-ICP-MS at uOttawa | | | EPMA | |
| | LA-ICP-MS | EPMA | | | | | |
| Isotope | 52μm | 20μm | Working Value | Average (n=30) | St.Dev | Average (n=10) | St.Dev |
|---|---|---|---|---|---|---|---|
| Mg24 | 0.1 | 89 | 10860 | 11441 | 636 | 13739 | 4560 |
| Al27 | 0.2 | 62 | 19437 | 18351 | 1021 | 22318 | 9961 |
| Si29 | 42 | 55 | 220 | 580 | 292 | 234 | 55 |
| Ca44 | 16 | 65 | 24 | 11 | 25 | 50 | 46 |
| Sc45 | 0.03 | - | 29 | 25 | 1 | - | - |
| Ti47 | 0.1 | 198 | 81967 | 77831 | 3757 | 77774 | 2103 |
| V51 | 0.01 | 90 | 9059 | 9212 | 277 | 7237 | 232 |
| Cr52 | 0.1 | 170 | 1096 | 1371 | 39 | 1143 | 106 |
| Mn55 | 0.06 | 106 | 1987 | 1987 | 101 | 1983 | 148 |
| Co59 | 0.01 | - | 225 | 294 | 11 | - | - |
| Ni60 | 0.8 | 116 | 536 | 599 | 18 | 589 | 72 |
| Cu63 | 0.03 | - | 31 | 7 | 18 | - | - |
| Zn66 | 0.1 | 525 | 500 | 388 | 28 | 362 | 370 |
| Ga69 | 0.02 | - | 41 | 48 | 2 | - | - |
| Ge74 | 0.05 | - | 0.9 | 0.7 | 0.06 | - | - |
| Y89 | 0.001 | - | 0.08 | 0.01 | 0.01 | - | - |
| Zr90 | 0.01 | - | 28 | 20 | 2 | - | - |
| Nb93 | 0.002 | - | 2 | 1 | 0.06 | - | - |
| Mo95 | 0.01 | - | 0.8 | 0.4 | 0.03 | - | - |
| Sn118 | 0.29 | - | 2 | 2 | 0.6 | - | - |
| Hf178 | 0.003 | - | 0.6 | 0.8 | 0.05 | - | - |
| Ta181 | 0.002 | - | 0.07 | 0.1 | 0.01 | - | - |
| W182 | 0.006 | - | 0.5 | 0.02 | 0.01 | - | - |
| Pb208 | 0.01 | - | 2 | 0.7 | 0.2 | - | - |

[57]Fe was used as the internal standard using a value of 57.2 wt.%. Working values (whole-rock by INAA –in bold) and informational values (by LA-ICP-MS in italics), taken from Dare et al. (2014).

6.2    Results

Within all Fe-oxide phases, elements which regularly have concentrations above detection limits are Mg, Al, Sc, Ti, V, Cr, Mn, Co, Ni, Zn, Ga, Ge, and Zr. Analyses of Sn can intermittently display values above detection limits, but, accuracy and precision cannot be assured with concentrations <3ppm based on analysis of reference materials. Ilmenite is the only Fe-oxide phase where Hf, Nb, and Ta regularly display concentrations above detection limits. Within Al-spinel Sn, W, and Pb occur in concentrations above detection limits but, cannot be considered accurate and precise due to concentrations approximately less than 1 order of magnitude higher than their respective detection limits. The concentration of Cu is highly variable possibly due to sub-microscopic inclusions of sulfide minerals that were not obvious during data reduction. Platinum group elements (PGE), rare-earth elements (La, Sm, Yb and Y), and precious metals (Ag and Au) were not detected during analysis. Full results of LA-ICP-MS analyses of Fe-oxides are provided in Appendix 5.

6.2.1    Comparison of LA-ICP-MS and EMPA data

The purpose of comparing analyses of magnetite obtained by LA-ICP-MS is to 1) check if using the Fe value from EMPA is an appropriate method of data reduction, 2) determine the effect of incorporating exsolutions into the LA-ICP-MS analyses, and 3) verify calibration of LA-ICP-MS using reference materials described in Section 6.1. Analyses of magnetite, in both massive and disseminated samples, by EMPA and LA-ICP-MS are generally in agreement for most elements (Ti, Mg, Mn, Ni, and Cr) when concentrations are above detection limits given by EMPA (Fig. 6.2). This indicates that the Fe value used to treat the LA-ICP-MS data is correct. However, analyses of Ti within a few magnetite from disseminated samples are significantly higher by LA-ICP-MS relative to EMPA. Within magnetite from disseminated samples, coarser exsolutions of ilmenite are more abundant (Fig.

6.1A) and lack the relatively even distribution compared to magnetite from massive samples. Therefore, the smaller beam size used for EMPA analyses is less likely to incorporate similar quantities of ilmenite exsolutions compared to LA-ICP-MS analysis causing the lower Ti contents by EMPA. Analyses of V within magnetite by EMPA and LA-ICP-MS show a >10% disagreement in massive and disseminated samples (Fig 6.2). As shown by analyses of in-house standard BC-28 by probe and laser, V is also significantly underestimated by approximately 20 % (relative difference) by EMPA, probably due to overcorrecting the interference of Ti on V. Therefore LA-ICP-MS data of V is preferred.

For ilmenite, the only elements in agreement by both EMPA and LA-ICP-MS methods are Mg, Mn (Fig. 6.3) and Cr, when above detection limit of the probe. Even though ilmenite contains concentrations well above EMPA detection limits for Ti and Al, concentrations of these elements are significantly higher in EMPA analyses relative to LA-ICP-MS analyses. The presence of Al-spinel inclusions/exsolutions within ilmenite has been noted by petrography which could generate heterogeneous distributions of Al within the analyzed crystal. There are several potential explanations for the disagreement in Ti contents. Exsolution phases within ilmenite included during LA-ICP-MS analyses could be an obvious explanation but, are only visible within hemo-ilmenite from Lac à l'Orignal (Fig. 6.1B). The small beam diameter used during EMPA analysis does not permit adequate incorporation of hematite exsolutions within a primary ilmenite crystal. However, LA-ICP-MS analyses incorporates both primary ilmenite host and hematite exsolutions into the analysis. This would result in the increased incorporation of Fe into the analysis which could cause the decrease in Ti and Mn contents shown in Figure 6.3. For ilmenite that does not show evident exsolutions, the Fe value from the probe is valid because all the other minor elements agree between the two methods, including Mn. The effect of using a reference material for calibration with low Ti concentrations (450 ppm) relative to ilmenite (>43% wt.% $TiO_2$; Appendix 5) is unknown.

Comparison of EMPA and LA-ICP-MS analyses of Al-spinel are displayed in Fig. 6.4. Elements that display concentrations well above detection limits within Al-spinel (e.g., Mg, Al, Cr and Zn) have EMPA and LA-ICP-MS values in agreement with each other. It is important to note that Al-spinel is the only Fe-oxide phase with relatively equal Al values obtained from both EMPA and LA-ICP-MS analyses. Analyses of V in Al-spinel by EMPA likely disagree with those of LA-ICP-MS for similar reasons to that of magnetite. There are two possible explanations for the disagreement of Ti in Al-spinel by the two methods; 1) some of the values are near or below detection limit or 2) LA-ICP-MS analysis incorporated possible exsolution phases that were not noted within coarse Al-spinel crystals by petrography. The effect of low Ti contents of the reference material GSE-1g would be negligible here because maximum Ti contents in Al-spinel are approximately 2500 ppm.

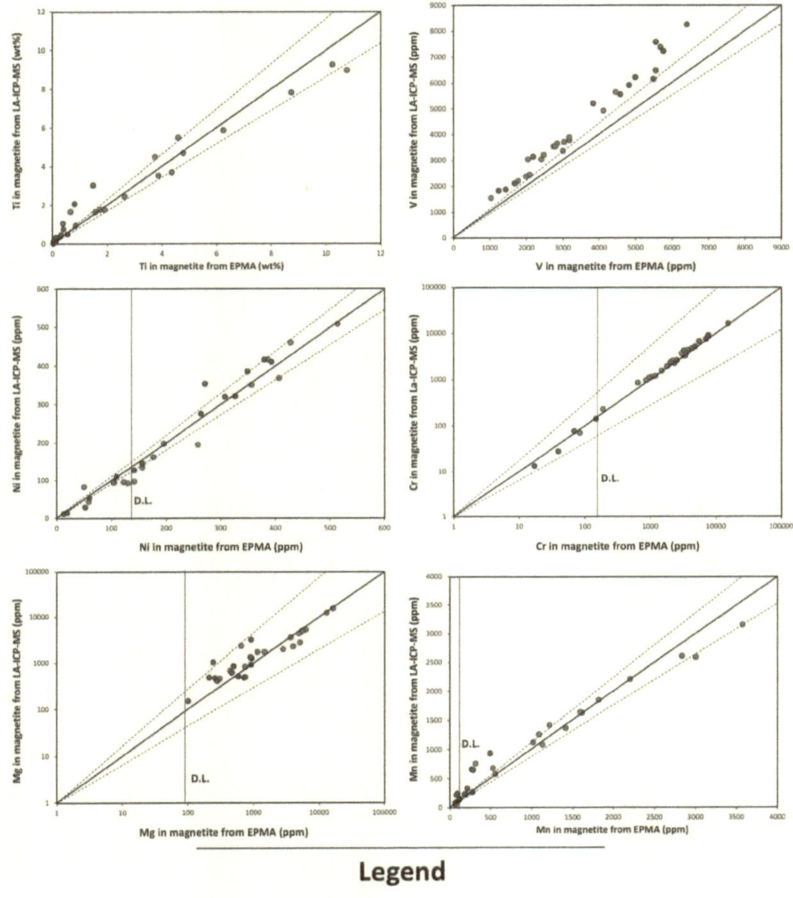



**Figure 6.2:** Comparison of average magnetite compositions as determined by LA-ICP-MS and EMPA for elements that were analyzed by both techniques. Red circles indicate analyses of magnetite that came from samples of massive ore and blue circles indicate analyses of magnetite which came from samples containing disseminated Fe-oxides. Thick solid line represents a 1:1 slope. Dotted lines represent 10% differences from the 1:1 slope. Solid vertical lines indicate the detection limit of an element by EMPA. Al and Zn are not plotted on this figure due to the immense scatter of Al concentration due to exsolutions and Zn values are below detection limit by EMPA.

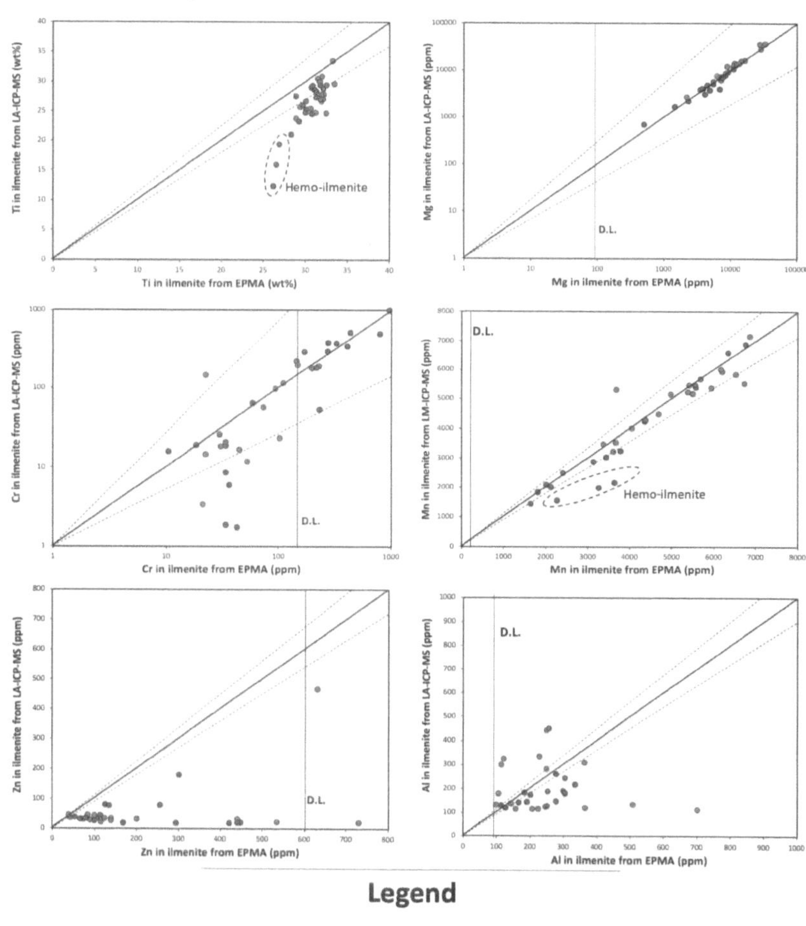



## Legend

● Massive/semi-massive sample

● Disseminated sample

**Figure 6.3:** Comparison of average ilmenite compositions as determined by LA-ICP-MS and EMPA for elements that were analyzed by both techniques. Red circles indicate analyses of ilmenite that came from samples of massive ore and blue circles indicate analyses of ilmenite which came from samples containing disseminated Fe-oxides. Thick solid line represents a 1:1 slope. Dotted lines represent 10% differences from the 1:1 slope. Solid vertical lines indicate the detection limit of an element by EMPA. Ni and V contents are below detection limit in ilmenite by EMPA. Aluminium and Zn are not determined well by EMPA.

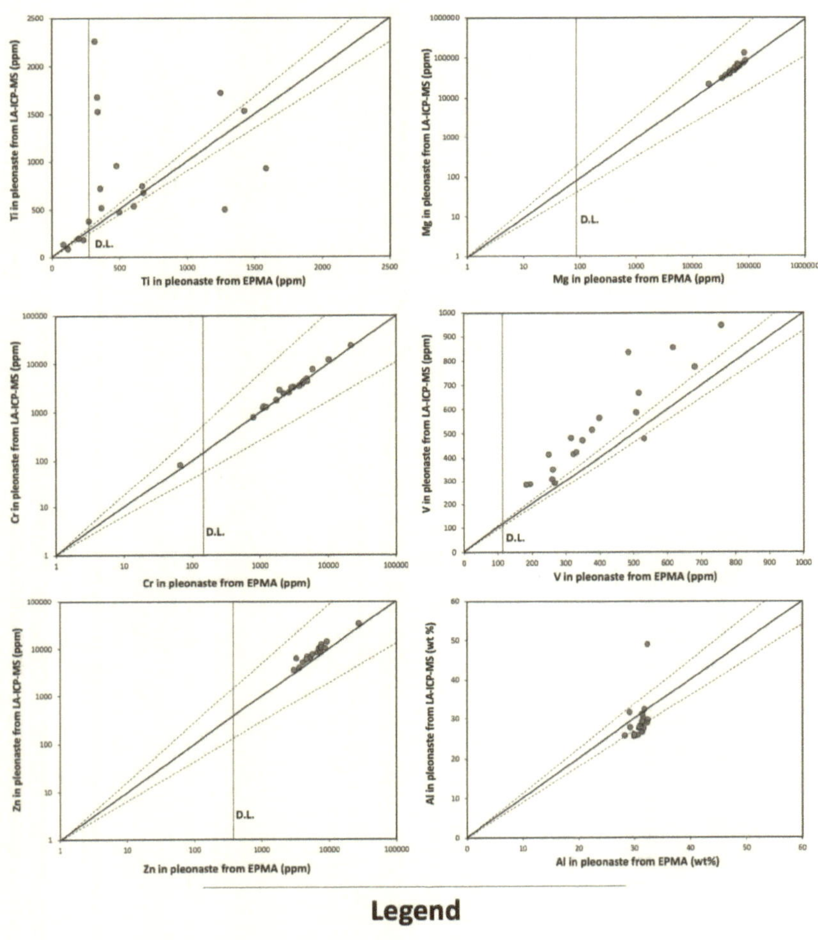



**Figure 6.4:** Comparison of average Al-spinel compositions as determined by LA-ICP-MS and EMPA for elements that were analyzed by both techniques. Red circles indicate analyses of Al-spinel that came from samples of massive ore, Al-spinel is only present in trace amounts as very fine-grains within massive samples. Thick solid line represents a 1:1 slope. Dotted lines represent 10% differences from the 1:1 slope. Solid vertical lines indicate the detection limit of an element by EMPA.

106

6.2.2    Distribution of trace elements among the Fe-oxide phases

Before understanding the behavior of trace elements in magnetite during magmatic and post-magmatic processes, the distribution of trace elements between magnetite and co-existing ilmenite and Al-spinel must be constrained. It is already known that ilmenite is significantly enriched in Ti, Mg, Mn, W and other HFSE's (Sc, Nb, Ta, Zr, Hf) relative to magnetite (e.g., Dare et al. 2012, 2014; Arguin et al. 2018). Therefore, magnetite which would 1) crystallize after, 2) co-crystallize with abundant ilmenite, or 3) exsolve significant ilmenite should be depleted in elements which preferentially partition into ilmenite, as has been noted in other anorthosite hosted Fe-oxide deposits (e.g., Lac a Paul Ti-P deposit-Néron 2012). However, none of the studies mentioned (Néron 2012; Dare et al. 2014) have considered the possible effect of significant Al-spinel exsolution from magnetite, which is seen in many of the samples examined in this study, on the primary composition of magnetite. Although Al-spinel is generally present in lower modal percentages relative to ilmenite in this study, effect of Al-spinel exsolution may be less significant than the effect of ilmenite oxy-exsolution on primary magnetite composition.

In order to constrain the distribution of trace elements among magnetite, ilmenite, and Al-spinel the following steps were taken; 1) average trace element compositions for magnetite, ilmenite, and Al-spinel for each deposit were calculated, 2) enrichment factors were obtained for magnetite/ilmenite (i.e. a two-mineral partition coefficient) by dividing the concentration of an element in magnetite by that in ilmenite ($Conc_{Mt}/Conc_{ilm}$), and 3) enrichment factors were obtained for magnetite/Al-spinel by dividing the concentration of an element in magnetite by that of Al-spinel ($Conc_{Mt}/Conc_{Spl}$). Results for magnetite/ilmenite and magnetite/Al-spinel partitioning behaviors are shown on Figs. 6.5 and 6.6, respectively. Analyses from Buttercup (Fe-Ti-V) were used as a baseline for comparison (Fig. 6.5A) since magnetite is the predominant Fe- oxide phase within the mineralized zone (>90% overall) and apatite is absent. This means that Buttercup contains magnetite which 1) likely underwent low amounts of compositional modification due to the presence of competing ilmenite and Al-spinel exsolution products and 2) crystallized from a relatively primitive magma, since apatite is not a cumulus mineral phase and it is relatively rich in V.

Magnetite/ilmenite partitioning behavior for Buttercup shows that magnetite is relatively enriched in Cr, Ni, V, Co, Zn, Ga, Mo, Ge, and Al whereas ilmenite is enriched in Ti, Mn, Mg, W and HFSEs (Fig. 6.5A), in agreement with previous work. The partitioning behavior of the trace elements between magnetite and ilmenite generally remains the same for all the deposits but the amount of enrichment changes significantly. Figure 6.5 shows that the degree of Mg, Mn, W and HFSE depletion is proportional to the degree of Ti depletion within magnetite. Since Ti is a primary component of ilmenite, this means that these elements can possibly become depleted within magnetite in the presence of ilmenite. From lowest to highest degree, the increasing presence of ilmenite has modified the composition of magnetite as follows: 1) Buttercup (Fig. 6.5A), 2) St. Charles de Bourget (Fig. 6.5C), 3) Lac Perron (Fig. 6.5D), 4) Lac Margane/Houlière-Nord (Fig. 6.5B), 5) Lac à Paul (Fig. 6.5E), 6) Lac à l'Orignal (Fig. 6.5F). The depletion of Ti within magnetite can possibly be attributed to the modal percentage of coarse-grained ilmenite (Fig. 6.7).

The Buttercup Fe-Ti-V deposit (Fig. 6.6A) is also used as the baseline for comparing the partitioning behavior of trace elements between magnetite and Al-spinel. Fig. 6.6 shows that magnetite is generally more enriched in HFSE, Cu, Mo, V, and Cr whereas Al-spinel is strongly enriched in Zn, Mg, Ge, and Al and weakly enriched in Ni, Co, and Ga. This pattern remains relatively similar for most deposits/occurrences but, does deviate slightly for samples from Lac Margane/Houlière-Nord (Fig. 6.6B), Lac à Paul (Fig. 6.6E), and Lac à l'Orignal

(Fig. 6.6F) deposits/occurrences where all or some of the HFSE are unusually enriched in Al-

spinel relative to magnetite. This may be due to lower precision in the determination of HFSE

by LA-ICP-MS analysis of Al-spinel at concentrations <10 ppm and lower modal percentage

of Ti-magnetite relative to ilmenite. Although the partitioning behavior for elements between

magnetite and Al-spinel can deviate among the deposits, the degree of depletion for Zn and

Co appear to be proportional to depletions of Al and Mg. Since Al and Mg are primary

components of Al-spinel, this means that these elements become depleted within magnetite

due to the presence of Al-spinel. As Mg is also preferentially partitioned into ilmenite relative

to magnetite (Fig. 6.5) it is not as reliable to determine the effects of co-existing Al-spinel on

magnetite composition. From lowest to highest degree, the presence of Al-spinel has modified

the composition of magnetite from 1) Buttercup (Fig. 6.7A, 2) St. Charles de Bourget (Fig.

6.7C), 3) Lac Perron (Fig. 6.7D), 4) Lac à Paul (Fig. 6.7E), 5) Lac Margane/Houlière-Nord

(Fig. 6.7B), 6) Lac à l'Orignal (Fig. 6.7F). In summary, the partitioning behaviors shown in

Figs. 6.5 and 6.6 show that 1) V, Cr, and Mo (except for Lac à l'Orignal) preferentially

partition into magnetite, 2) Ti, Mn, Sc, and HFSE preferentially partition into ilmenite, 3) Al,

Co and Zn preferentially partition into Al-spinel, and 4) Mg partitions into both ilmenite and

Al-spinel over magnetite.

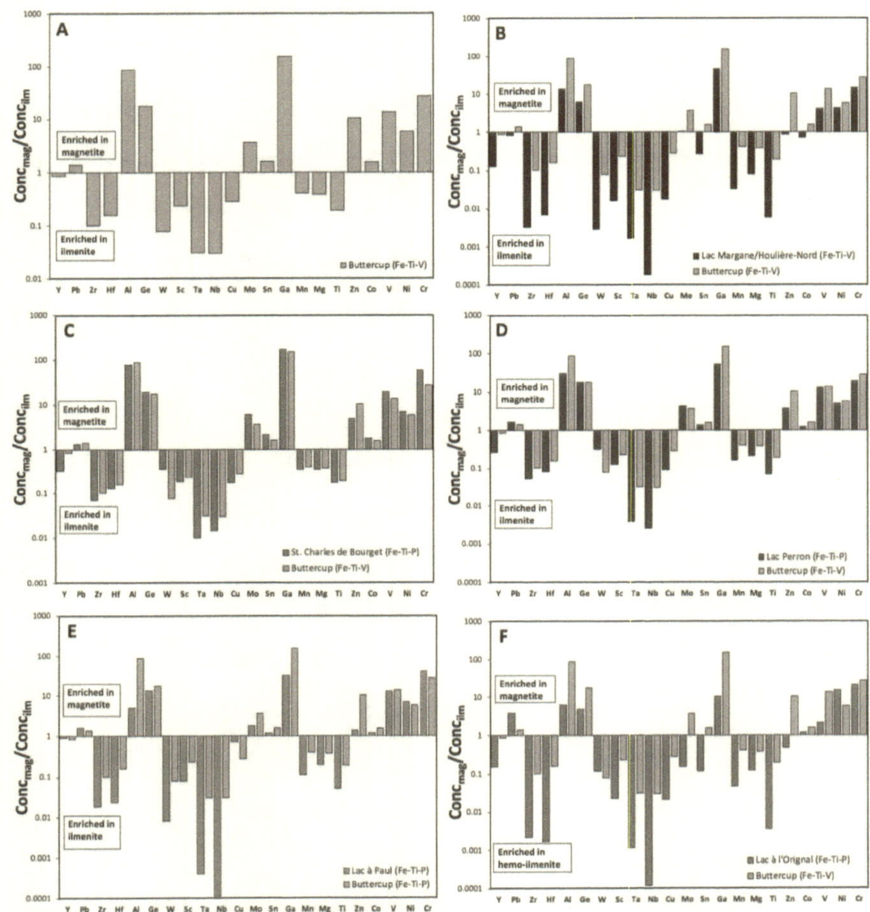

**Figure 6.5:** Bar graphs displaying the partitioning behavior of trace elements between magnetite and ilmenite in the Fe-Ti-V-P mineralization from Lac St. Jean area, as quantified by an enrichment factor (Conc$_{Mag}$/Conc$_{Ilm}$). A) shows values obtained from Buttercup which are then compared to partitioning behavior from B) Lac Margane/Houlière-Nord, C) St. Charles de Bourget, D) Lac Perron, E) Lac à Paul, and F) Lac à l'Orignal. Mag = magnetite, ilm = ilmenite.

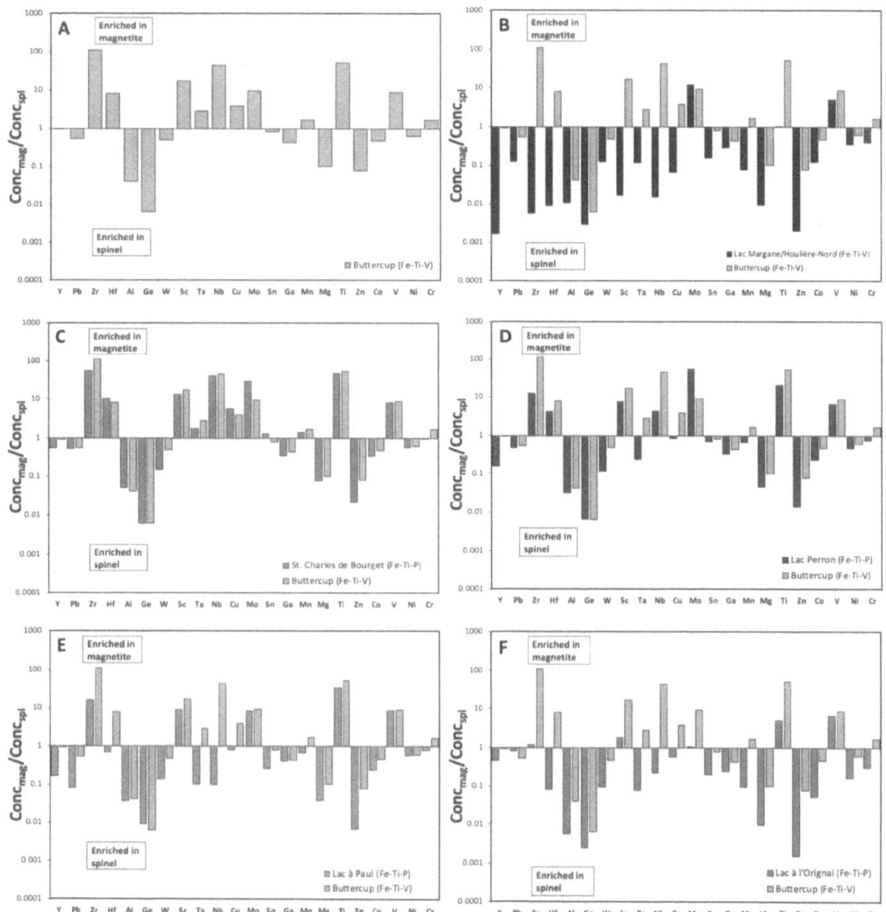

**Figure 6.6:** Bar graphs displaying the partitioning behavior of trace elements between magnetite and spinel (Al-spinel), as quantified by an enrichment factor ($Conc_{Mag}/Conc_{spl}$). A) shows values obtained from the Buttercup deposit which are then compared to partitioning behavior from B) Lac Margane/Houlière-nord, C) St. Charles de Bourget, D) Lac Perron, E) Lac à Paul, and F) Lac à l'Orignal with those of Buttercup. Mag = magnetite, spl = spinel.

111

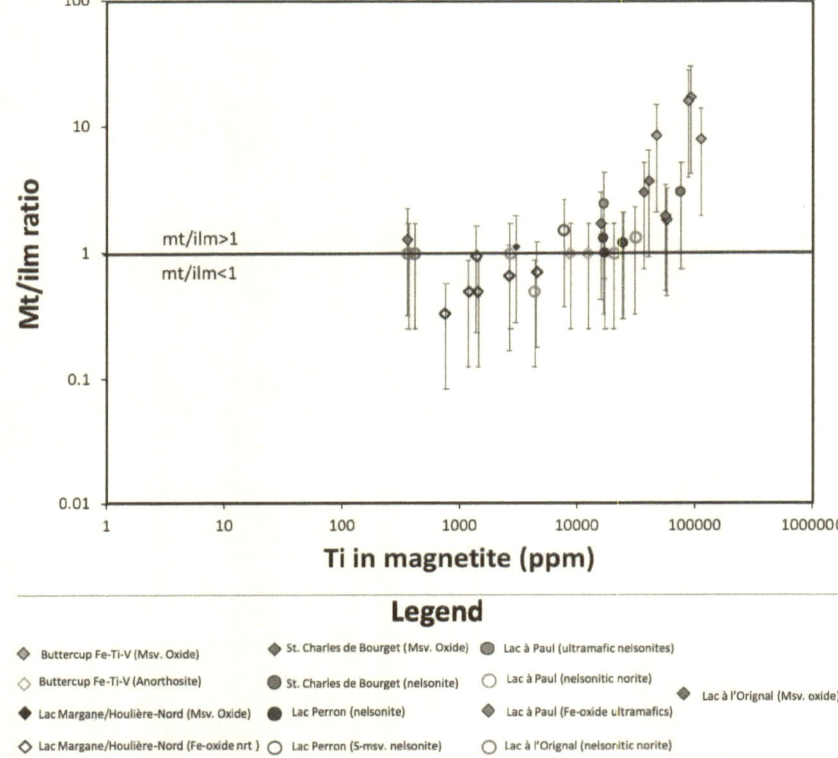


**Figure 6.7:** Binary plots displaying the effect of co-existing ilmenite on magnetite compositions. Magnetite (Mt)/Ilmenite (ilm) ratios, obtained by petrographic observations, vs. Ti content of magnetite. Error bars indicate standard deviation. Data for St. Charles de Bourget supplemented by Dare et al. (2014) and data for Lac à Paul supplemented by Néron (2012). Msv=massive, nrt=norite, s-msv=semi massive.

6.2.3   Trace element variation in magnetite from the Fe-Ti-V-P deposits of the Lac St. Jean anorthosite suite

Plotting magnetite LA-ICP-MS trace element data on the multi-element diagram created by Dare et al. (2014) permits the discrimination of magnetite from different ore forming environments. This includes Fe-Ti-V and Fe-Ti-P deposits, both of which occur within the Lac St. Jean anorthosite suite. Elements are normalized to bulk continental crust (normalization values fromRudnick and Gao 2003) and arranged from left to right in order of increasing compatibility of magnetite. Dare et al. (2014) chose to normalize these elements to bulk continental crust because; 1) bulk continental crust is closer in composition to an evolved silicate melt that would crystallize magnetite (e.g. ferrodiorite) and 2) hydrothermal fluids are more likely to have interacted with crustal rocks when compared to a primitive mantle source. Dare et al. (2014) noted that magnetite from Fe-Ti-V deposits (defined by the economic magnetite layers of the Bushveld Complex) was enriched in Mg, Ni, Co, V, and Cr relative to magnetite from Fe-Ti-P deposits indicating crystallization from a more primitive magma.

However, geochemically discriminating magnetite from Fe-Ti-V and Fe-Ti-P deposits is best performed using a small number of elements (Cr, Ni, V $\pm$ Ti, Mo, Ga, Nb, and Ta; Dare et al. 2014). Since the partitioning behavior of elements between magnetite and co-existing ilmenite and Al-spinel is now better constrained for the Fe-Ti-V-P deposits/occurrences from the Lac St. Jean anorthosite suite, it is now possible to evaluate the effect of the presence of granular ilmenite and Al-spinel on magnetite composition: if granular ilmenite is present, the co-existing magnetite most likely be depleted in Ti, Nb, and Ta and if granular Al-spinel is present, Al, Zn, and Co contents of magnetite will likely be depleted too, as shown by Figs. 6.5 and 6.6. Knowing this, Cr, Ni, V, and Mo contents of magnetite are the likely best potential candidates for determining the degree of magma evolution from which the oxides crystallized.

The multi-element diagrams in Figure 6.8 show the variability of magnetite composition from the different oxide deposits/occurrences in the Lac St Jean area. Similar to Figs. 6.5 and 6.6, magnetite from the Buttercup deposit will be used as a baseline to compare the ore-forming environments among the deposits/occurrences examined in this study because the massive oxides contain the least amount of granular ilmenite and Al-spinel. Magnetite from lense A (massive oxide) of the Buttercup deposit plots within the Fe-Ti-V field for all elements except for V, which is depleted (V = 3537-4319 ppm) relative to the Fe-Ti-V deposit field (Fig. 6.8A). Magnetite within massive oxide from lense C of the Buttercup deposit also plots within the Fe-Ti-V deposit field but is depleted in Ti, Mg, Zn, and HFSE relative to magnetite from lense A. Disseminated magnetite hosted within the anorthosite near the contacts with the ore lenses at Buttercup shows strong depletions in Ti, Mg, Mn, and HFSE but is enriched in Cr and Zn relative to magnetite from massive ores from both lenses A and C. In general, there are 3 obvious trends shown in Fig. 6.8. First, magnetite from massive oxide samples/nelsonite from Lac Margane/Houlière-Nord (Fig. 6.8B), Lac Perron (Fig. 6.8D), and Lac à Paul (Fig. 6.8E) which contain higher proportions of granular ilmenite ± spinel is more depleted in Ti, Mg, Mn, Sc, HFSE (elements that partition into ilmenite) along with Zn and Co (elements that partition into spinel) and subsequently enriched in Cr, V, Mo, Ga, Ge (elements that partition into magnetite) relative to magnetite from Buttercup massive oxides. Depletions in Al related to the presence of Al-spinel within massive samples, however, are not obvious from Fig 6.8. Second, magnetite from nelsonite samples is enriched in Mo, Sn, and HFSE and depleted in V, Ni, and Cr relative to massive oxide samples Finally.

magnetite from disseminated samples is depleted in Ti, Mg, Al, and HFSE and enriched in Cr relative to magnetite from massive oxide samples from the same location.

The only deviations for these general trends described above are 1) nelsonite from Lac Perron contains magnetite that is enriched in Cr relative to magnetite from Buttercup massive ore (Fig 6.8D); 2) at Lac à Paul where Cr is depleted within magnetite from disseminated samples (nelsonitic norite) relative to semi-massive samples (ultramafic nelsonites; Fig 6.8E), 3) at Lac à l'Orignal where magnetite from all lithologies is significantly depleted in Mo relative to the Buttercup field (Fig. 6.8F), and 4) also at Lac à l'Orignal where magnetite from nelsonitic norites (disseminated) is depleted in Cr relative to magnetite from massive oxide (Fig 6.8F). The concentration of V does not appear to change significantly from massive to disseminated samples on the deposit scale.

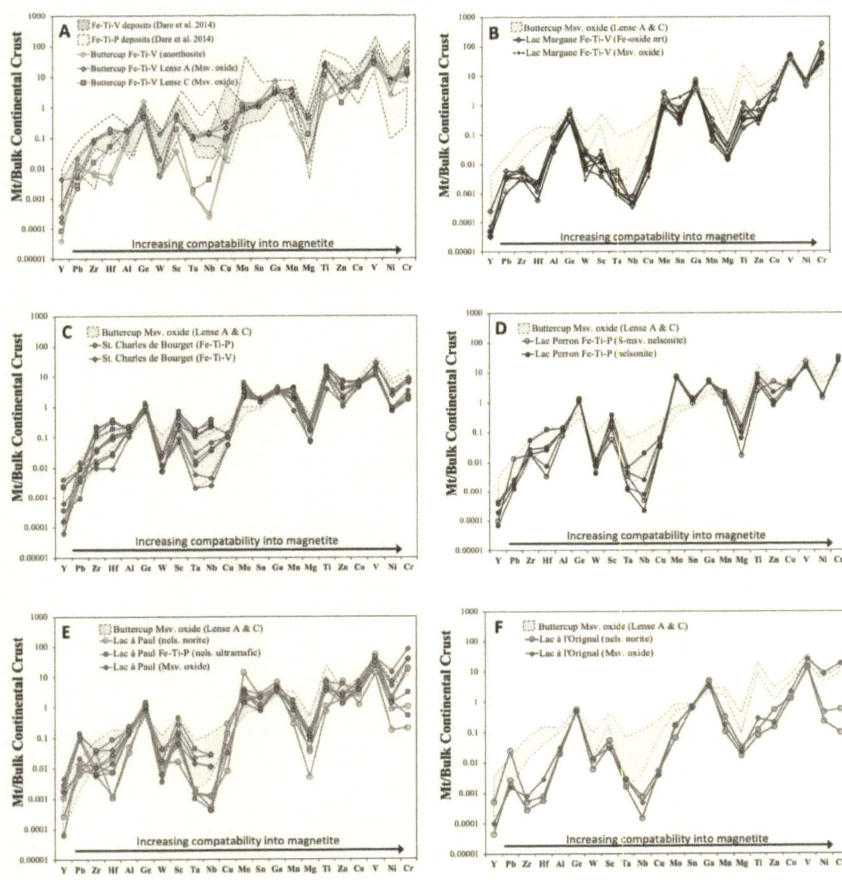

**Figure 6.8:** Multi-element variation diagrams for magnetite from Fe-Ti-V-P mineralization of Lac St. Jean area (after Dare et al. 2014). A) Buttercup Fe-Ti-V deposit, compared with ranges obtained by Dare et al. (2014) of magnetite composition from Fe-Ti-V (Bushveld Complex, S. Africa) and Fe-Ti-P deposits (Sept Iles, Canada; St. Charles de Bourget, Canada; Bushveld Complex). B) Lac Margane/Houlière-Nord Fe-Ti-V occurrences, C) St. Charles de Bourget Fe-Ti-V and Fe-Ti-P occurrences (data supplemented by Dare et al. 2014), D) Lac Perron, Fe-Ti-P occurrence, E) Lac à Paul (Fe-Ti-P) New Zone (data supplemented by Néron (2012) from Lac à Paul region) and F) Lac à l'Orignal Fe-Ti-P occurrence. The range of values from Buttercup (blue field) is shown for comparison in B-F. Each line plotted represents the average abundance of an element within magnetite normalized to bulk continental crust (values fromRudnick and Gao, 2003). Mt = magnetite, Msv = massive, nrt = norite, S-msv = semi- massive, nels = nelsonitic.

6.2.4    Evaluation of fractionation trends among mineralization

Previous reviews on magmatic oxide deposits hosted in anorthosites and layered intrusions indicate that Fe-Ti-V deposits represent oxide mineralization at its most primitive stage whereas Fe-Ti-P deposits formed from more evolved magma (Hébert et al. 2005; Woodruff et al. 2013; Dare et al. 2014; Charlier et al. 2015). This can be illustrated in Figs. 6.9A-B which shows that magnetite from Fe-Ti-V deposits (and massive oxide samples) is generally enriched in Ni, V, and Cr relative to magnetite from Fe-Ti-P deposits. Magnetite from the Lac Perron Fe-Ti-P deposit contains anomalously high Cr relative to other Fe-Ti-P deposits but is depleted in Ni and V relative to magnetite from Fe-Ti-V deposits. Fractionation trends are not as obviously displayed within Figs 6.9C-D. However, magnetite from disseminated samples generally contains lower Co (Fig. 6.9C) and Mg (Fig. 6.9D) contents relative to magnetite from massive/semi-massive samples. Depletions in Co and Mg within magnetite could possibly be related to the presence of Al-spinel and/or Fe-Mg silicates. Although Co contents within magnetite (Fig. 6.9C) appear to be less susceptible to depletions from massive to disseminated samples relative to Mg (Fig. 6.9D). A slight negative correlation is seen between Mo and Cr with the exception of Lac à l'Orignal, where magnetite has significantly lower Mo contents (Fig. 6.9E). Concentrations of Ga appear to remain relatively constant within magnetite from both Fe-Ti-V and Fe-Ti-P deposits and massive and disseminated samples (Fig. 6.9F).

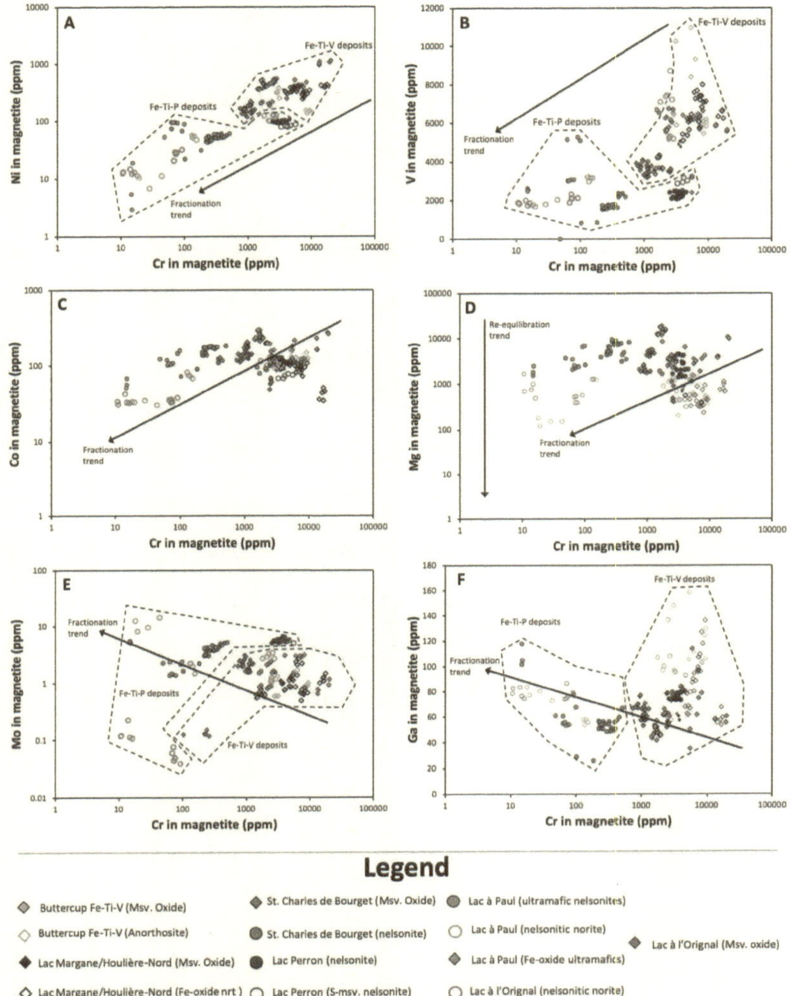

**Figure 6.9:** Binary diagrams displaying concentrations of elements in magnetite that are compatible (A-D) and incompatible (E-F) during fractionation plotted against Cr in magnetite, used as a proxy for the fractionation or re-equilibration (with Fe oxides or Fe-Mg silicates) trend as shown by labeled arrow(s). A) Ni vs. Cr, B) V vs. Cr, C) Co vs. Cr, D) Mg vs. Cr -downwards arrow represents re-equilibration with silicate trend caused by decreasing modal % of Fe-oxides, E) Mo vs. Cr and F) Ga vs. Cr. In A and B, magnetite from Fe-Ti-V an Fe-Ti-P deposits are separated by polygons. Data for St. Charles de Bourget supplemented by Martin-Tanguay (2012) and data for Lac à Paul deposit supplemented by Néron (2012). Msv = massive, g/n = gabbro/norite, S-msv = nelsonite.

# Chapter 7 Trace element chemistry of apatite

This section describes analytical methods (including comparison of LA-ICP-MS data with EMPA data), and concentrations of, and correlations between, trace elements within apatite with a particular focus in REE. Apatite grains were analyzed from the 4 Fe-Ti-P deposits/occurrences: 1) St. Charles de Bourget, 2) Lac Perron, 3) Lac à Paul, and 4) Lac à l'Orignal. Another important note is that apatite is present within 3 different lithologies in this study; 1) nelsonite (from Lac Perron and St. Charles de Bourget), 2) ultramafic nelsonite (dunitic nelsonite and peridotitic nelsonite from Lac à Paul), and 3) nelsonitic norite (from Lac à Paul and Lac à l'Orignal). Therefore, the apatite results displayed in this section will be interpreted based on 1) the sampling location (mineralization from which the apatite was sampled and location within the mineralized zone) and 2) lithology from which the apatite is hosted. Overall, the results displayed in this section aim to compare the compositions of apatite from the different Fe-Ti-P deposits examined in this study.

## 7.1 Analytical Methods

LA-ICP-MS analysis took place at the University of Ottawa using a Photon Machines Analyte (193 nm) Excimer laser system and 7700 Agilent Quadropole ICP-MS; analytical settings are shown in Tables 7.1 and 7.2. The analytical procedure involved the ablation of approximately 300 μm lines within an apatite crystal (30 s gas blank, 60 s signal with 5 μm/s stage speed), using a beam diameter of 52 μm. However, for smaller apatite grains from Lac à Paul New Zone samples (SDLP-37 and SDLP-38) the beam diameter and the stage speed were reduced to 25 μm and 4 μm/s respectively. Inclusions of Fe-oxides were avoided during the ablation process and data reduction. Calibration was performed using NIST-610, a synthetic glass doped with approximately 450 ppm of REE (rare earth elements) (Jochum et al. 2011). Monitoring of data quality was performed using GSE-1g, NIST-612, (synthetic glass doped with approximately 35 ppm of each REE) (Jochum et al. 2011), and in-house reference material Durango apatite. Analysis of 8 unknowns (apatite grains) was both preceded and proceeded by analysis of certified reference materials and in-house standards. Analyses of certified reference materials and in-house reference materials are displayed in Tables 7.3 and 7.4 respectively. Data reduction was performed by Glitter® using $Ca^{42}$ obtained by EMPA analysis of apatite as the internal standard. Elements such as Si, Al, Ti, Fe and K were used to monitor for inclusions during data reduction via Glitter®.

Table 7.3 shows that repeated analysis of GSE-1g produced accurate and precise results, because all elements have relative differences at or below 5% and have relative standard deviations of 5% or less. Analyses of NIST-612 also produced accurate and precise results for all elements except Mg. This could possibly be due to high relative uncertainties (at 95% confidence level) with Mg in NIST series of reference materials (NIST-610-617) when compared to other lithophile elements (Jochum et al. 2011). Analyses of both reference materials used do not produce accurate or precise results for P as relative differences and relative standard deviations are >10%. LA-ICP-MS analyses of NIST reference materials performed by Jochum et al. (2011) showed a >20% uncertainty at 95% confidence level with P indicating that analyses cannot be determined accurately and precisely. In any case, P is best determined by EMPA rather than LA-ICP-MS.

Relative differences and relative standard deviations of <20% will be accepted as accurate and precise for Durango apatite since it is a natural material and thus, heterogeneous relative to certified reference materials. When compared with working values of the same Durango apatite grain (DRC: EMPA values from Desormiers 2015; solution analysis by Dare unpub.), analyses of DRC at the University of Ottawa from this study are considered accurate and precise more most elements (Table 7.4) except for Al, Si, and K and Cl. Other examples of Durango apatite are given in Table 7.4 (Smithsonian Durango published values from Chew et al. 2016) are similar but not identical to uOttawa Durango grain (DRC) This can be attributed to natural variation within different crystals of Durango apatite.

**Table 7.1:** Analytical settings of LA-ICP-MS analysis for apatite at the University of Ottawa.

| | University of Ottawa |
| --- | --- |
| **Laser Ablation System** | Photon-Machines Analyte 193nm Excimer |
| **ICP-MS** | Agilent 7700x |
| **Laser Frequency** | 15 Hz |
| **Pulse Energy** | 5 mJ/pulse |
| **Stage Speed** | 5 µm/s (52 µm beam diameter), 4 µm/s (25 µm beam diameter) |
| **Beam Diameter** | 52 µm, 25 µm |
| **Dwell Time** | See Table 2 |
| **Analysis Procedure** | 30 s of gas blank, 60 s of signal |
| **Carrier Gas Flow** | Helium (1/L/min) |
| **Ar Flow Rate** | 0.7 L/Min |
| **Internal Standard** | $Ca^{42}$ (data from EPMA) |
| **Reference Material for Calibration** | NIST-610 (NIST-synthetic glass) |
| **Reference Material for Monitoring** | GSE-1g (USGS-synthetic glass) |
| | NIST-612 (NIST-synthetic glass) |
| **In-house Monitor** | Durango (Fluoro-apatite from the Cerro de Mercado IOA deposit) |
| **Monitor Signal for Inclusions** | Si, Fe, Ti, Al, Cu |

**Table 7.2:** Isotopic dwell times used during LA-ICP-MS analysis at the University of Ottawa.

| Isotope | | Dwell Time (ms) |
| --- | --- | --- |
| 23 | Na | 5 |
| 25 | Mg | 10 |
| 27 | Al | 5 |
| 29 | Si | 5 |
| 39 | K | 8 |
| 55 | Mn | 8 |
| 88 | Sr | 10 |
| 89 | Y | 10 |
| 139 | La | 10 |
| 140 | Ce | 10 |
| 141 | Pr | 10 |
| 146 | Nd | 10 |
| 147 | Sm | 10 |
| 153 | Eu | 10 |
| 157 | Gd | 10 |
| 159 | Tb | 10 |
| 163 | Dy | 10 |
| 165 | Ho | 10 |
| 166 | Er | 10 |
| 169 | Tm | 10 |
| 172 | Yb | 10 |
| 175 | Lu | 10 |
| 232 | Th | 8 |
| 238 | U | 8 |

**Table 7.3:** Results of LA-ICP-MS analysis of reference materials used during the analysis of apatite. All element concentrations are given in ppm.

| Isotope | Detection Limits LA-ICP-MS 52-25µm | Calibration NIST610 Certificate Values Avg. | St.Dev | Monitor GSE-1g Certificate Values Avg. | St.Dev | uOttawa Avg. (n=10) | St.Dev | Monitor NIST612 Certificate Values Avg. | St.Dev | uOttawa Avg. (n=13) | St.Dev |
|---|---|---|---|---|---|---|---|---|---|---|---|
| Na23 | 2.5-18 | 99409 | 2226 | 28933 | 1484 | 30478 | 1444 | 101635 | 223 | 101464 | 1723 |
| Mg25 | 0.4-2.5 | 429 | 29 | 21106 | 181 | 19938 | 420 | 68 | 5 | 56 | 1 |
| Al27 | 0.2-1.6 | 10320 | 212 | 68804 | 2117 | 74923 | 971 | 10744 | 212 | 10500 | 203 |
| Si29 | 22-159 | 325806 | 2337 | 250994 | 7011 | 251163 | 7393 | 337024 | 2805 | 337900 | 10985 |
| P31 | 4-4.6 | 413 | 46 | 70 | 20 | 29 | 4 | 47 | 7 | 40 | 5 |
| Cl35 | 25-217 | 274 | 67 | 1330 | 130 | 1208 | 159 | 142 | 58 | 281 | 23 |
| K39 | 0.7-4.6 | 464 | 21 | 21800 | 200 | 23281 | 1169 | 62 | 2 | 67 | 2 |
| Mn55 | 0.05-0.29 | 444 | 13 | 590 | 20 | 606 | 22 | 39 | 1 | 38 | 1 |
| Sr88 | 0.004-0.2 | 516 | 1 | 447 | 5 | 464 | 4 | 78 | 0 | 79 | 2 |
| Y89 | 0.002-0.01 | 462 | 11 | 410 | 30 | 438 | 15 | 38 | 1.4 | 36 | 1.1 |
| La139 | 0.002-0.004 | 440 | 10 | 392 | 4 | 400 | 9 | 36 | 0.7 | 35 | 0.9 |
| Ce140 | 0.002-0.003 | 453 | 8 | 414 | 4 | 429 | 10 | 38 | 0.7 | 38 | 0.5 |
| Pr141 | 0.002-0.01 | 448 | 7 | 460 | 5 | 474 | 6 | 38 | 1.0 | 38 | 0.7 |
| Nd146 | 0.01-0.03 | 430 | 8 | 453 | 5 | 458 | 7 | 36 | 0.7 | 35 | 0.8 |
| Sm147 | 0.01-0.07 | 453 | 11 | 488 | 5 | 499 | 6 | 38 | 0.8 | 37 | 0.6 |
| Eu153 | 0.003-0.02 | 447 | 12 | 410 | 20 | 414 | 9 | 36 | 0.8 | 35 | 0.6 |
| Gd157 | 0.01-0.06 | 449 | 12 | 514 | 6 | 533 | 11 | 37 | 0.9 | 36 | 0.9 |
| Tb159 | 0.002-0.01 | 437 | 9 | 480 | 20 | 508 | 8 | 38 | 1.1 | 35 | 1.1 |
| Dy163 | 0.008-0.04 | 437 | 11 | 524 | 6 | 551 | 10 | 36 | 0.7 | 34 | 0.7 |
| Ho165 | 0.002-0.01 | 449 | 12 | 501 | 8 | 535 | 16 | 38 | 0.8 | 36 | 1.2 |
| Er166 | 0.006-0.03 | 455 | 14 | 595 | 6 | 577 | 28 | 38 | 0.9 | 36 | 0.7 |
| Tm169 | 0.002-0.01 | 435 | 10 | 500 | 20 | 543 | 18 | 37 | 0.6 | 35 | 1.0 |
| Yb172 | 0.008-0.05 | 450 | 9 | 520 | 5 | 541 | 12 | 39 | 0.9 | 37 | 0.8 |
| Lu175 | 0.002-0.01 | 439 | 8 | 518 | 6 | 553 | 18 | 37 | 0.9 | 35 | 1.2 |
| Th232 | 0.003-0.02 | 457 | 1 | 380 | 20 | 386 | 13 | 38 | 0.08 | 36 | 1.2 |
| U238 | 0.003-0.03 | 462 | 1 | 420 | 30 | 436 | 13 | 37 | 0.08 | 37 | 0.98 |

$Ca^{42}$ was used as the internal standard for all analyses using the following values: 8.15 wt.% (NIST-610), 5.29 wt.% (GSD-1g), and 8.50 wt.% (NIST-612). Certificate values for NIST reference materials are taken from Jochum et al. (2011) and certificate values for GSE-1g were taken from Jochum et al. (2005).

**Table 7.4:** Results of LA-ICP-MS and EMPA analysis of in-house reference material Durango apatite used to monitor the analysis of apatite. All element concentrations are given in ppm.

| | Detection Limits (ppm) | | In-house Monitor (DRC) Durango Apatite C (DRC) *Working Values | | | | | Durango Apatite C (DRC) [4]LA-ICP-MS (uOttawa) | | Other Durango apatite grains for comparison Smithsonian Durango [4]EPMA (uOttawa) | | Durango Literature Chew et al. 2016 |
| | LA-ICP-MS | EPMA | [1]EPMA | [2]Bulk | [3A]Bulk | [3B]Bulk | DRC | Average | | Average | | |
| Isotope | 52-25μm | 10μm | (n = 30) | analysis | analysis | analysis | Average | (n=14) | St.Dev | (n=4) | St.Dev | Average |
|---|---|---|---|---|---|---|---|---|---|---|---|---|
| Na23 | 2.5-18 | 185 | 1898 | 1706 | 1800 | 1800 | **1801** | 1579 | 144 | 2169 | 168 | n.a. |
| Mg25 | 0.4-2.5 | 102 | 111 | 181 | 100 | 100 | **123** | 103 | 5 | 94 | 63 | 83 |
| Al27 | 0.2-1.6 | 88 | BDL | 159 | <100 | <100 | **<100** | 1.04 | 0.47 | n.a. | n.a. | n.a. |
| Si29 | 22-159 | 97 | 1712 | 2197 | n.a. | n.a. | **1954** | 1923 | 262 | 1444 | 486 | n.a. |
| Cl35 | 25-217 | 100 | 3668 | 3670 | n.a. | n.a. | **3669** | 3369 | 686 | 4793 | 487 | n.a. |
| K39 | 0.7-4.6 | 139 | BDL | BDL | <100 | <100 | **<100** | 16 | 7 | 179 | 59 | n.a. |
| Mn55 | 0.05-0.29 | 405 | BDL | 100 | 88 | 90 | **93** | 96 | 8 | 100 | 109 | 95 |
| Sr88 | 0.004-0.2 | 551 | 511 | 480 | 567 | 576 | **533** | 500 | 38 | 210 | 83 | 502 |
| Y89 | 0.002-0.01 | 521 | BDL | 800 | >500 | >500 | **800** | 735 | 68 | 729 | 232 | 1017 |
| La139 | 0.002-0.004 | 1882 | 3359 | 3840 | 3540 | 3610 | **3587** | 3676 | 325 | 3732 | 1182 | 4239 |
| Ce140 | 0.002-0.003 | 1383 | 5913 | 5190 | >500 | >500 | **5551** | 5332 | 335 | 5382 | 970 | 5777 |
| Pr141 | 0.002-0.01 | - | - | 457 | 501 | 501 | **486** | 424 | 22 | - | - | 530 |
| Nd146 | 0.01-0.03 | - | - | 1485 | >1000 | >1000 | **1485** | 1349 | 116 | - | - | 1728 |
| Sm147 | 0.01-0.07 | - | - | 206 | 246 | 249 | **234** | 185 | 13 | - | - | 251 |
| Eu153 | 0.003-0.02 | - | - | 18.5 | 17.5 | 17.4 | **17.8** | 17.3 | 1.1 | - | - | 21 |
| Gd157 | 0.01-0.06 | - | - | 176 | 175 | 171 | **174** | 173 | 17 | - | - | 226 |
| Tb159 | 0.002-0.01 | - | - | 22.1 | 24.5 | 24.4 | **23.7** | 20.9 | 1.7 | - | - | 30 |
| Dy163 | 0.008-0.04 | - | - | 129 | 135 | 134 | **133** | 118 | 10 | - | - | 167 |
| Ho165 | 0.002-0.01 | - | - | 24.6 | 27.4 | 27.1 | **26.4** | 23.6 | 2.1 | - | - | 35 |
| Er166 | 0.006-0.03 | - | - | 70.3 | 76.0 | 74.1 | **73.5** | 63.9 | 4.5 | - | - | 92 |
| Tm169 | 0.002-0.01 | - | - | 8.4 | 8.5 | 8.2 | **8.4** | 7.9 | 1 | - | - | 12 |
| Yb172 | 0.008-0.05 | - | - | 45.7 | 48.1 | 43.5 | **45.8** | 41.8 | 2.8 | - | - | 60 |
| Lu175 | 0.002-0.01 | - | - | 5.1 | 5.2 | 5.2 | **5.2** | 5.0 | 0.4 | - | - | 6.6 |
| Th232 | 0.003-0.02 | - | - | 286 | 294 | 296 | **292** | 241 | 20 | - | - | 359 |
| U238 | 0.003-0.03 | - | - | 14.3 | 15.5 | 15.2 | **15** | 13.8 | 1.6 | - | - | 19 |

*Working values: 1) Electron microprobe analysis (EMPA) at uOttawa (Desormiers, 2015); 2-3) Bulk analysis (Dare unpublished) of DRC powder, from the same 1 x 4 cm gem crystal as laser mount, carried at ALS Chemex (Vancouver), by 2) XRF (majors) and ICP-MS (traces) after lithium metaborate fusions and 3) 4-acid digestion-ICP-MS. 4) This study. Abbreviations: n.a. not analysed, BDL. below detection limit.

## 7.2  LA-ICP-MS results

In total 77 grains (n) of apatite were analyzed from 9 apatite-rich samples (N) from St.
Charles de Bourget (N = 1, n = 7), Lac Perron (N = 2, n = 29), Lac à Paul New Zone (N = 3, n
= 16), Lac à l'Orignal (N = 3, n = 23). When considering lithology, the dataset comprises 3
samples of nelsonite (n = 36), 2 samples of ultramafic nelsonite (n = 10), and 4 samples of
nelsonitic norite (n = 29). Supplemental data for St. Charles de Bourget (averages for 2
samples) and Lac à Paul ore zones (averages for 36 samples) was provided by Desormiers
(2015) and Chartier-Montreuil (2017), respectively. Full results of analyses and detections
limits of elements analyzed in apatite by LA-ICP-MS are presented in Appendix 6. Elements
that consistently display concentrations above detection limits are Cl, Na, Mg, Mn, Sr, Y, all
REE, Th and U. Elements that were intermittently detected were Si, Al, and K. This is likely
due to the presence of inclusions that could not be detected during data reduction. Data
collected from the Lac à Paul New Zone appeared to display signs of significant
contamination as analyses indicated elevated As and Ti contents (Appendix 7), therefore,
these samples were omitted from graphing. This could possibly be related to the very fine
grain size of the apatite crystals at the Lac à Paul New Zone (Figs 4.13D - F).

### 7.2.1  Comparison with EMPA data

Comparison of apatite analyses by EMPA and LA-ICP-MS are shown in Figure 7.1.
Elements that agree to within 10% between the two analytical methods are Mg, Na, and Cl.
Elements which do not correlate well by probe and laser are P, Sr, and Mn. The poor
correlation of P between EMPA and LA-ICP-MS can possibly be explained by 1) relatively
poor accuracy and precision of P obtained from LA-ICP-MS compared to other elements
shown in this chapter, as shown by analyses of NIST reference materials (Jochum et al. 2011)
and 2) unknown effects of low P concentrations (413 ppm) within the reference material used

for calibration (NIST610). Detection limits for both Sr and Mn by EMPA were relatively high (400-550 ppm) with the analytical settings used during EMPA analysis, which was optimized for analysis of major elements within apatite and volatiles. As shown with the Fe-oxides, elemental concentrations within a mineral from EMPA that are reported below, at, or near detection limit by EMPA cannot be considered accurate or precise as they cannot be fully quantified. Detection limits by EMPA for La, Ce and Y (not plotted in Figure 7.1) were much too high to produce accurate and precise analysis of these elements in apatite by EMPA from the Fe-Ti-P deposits.

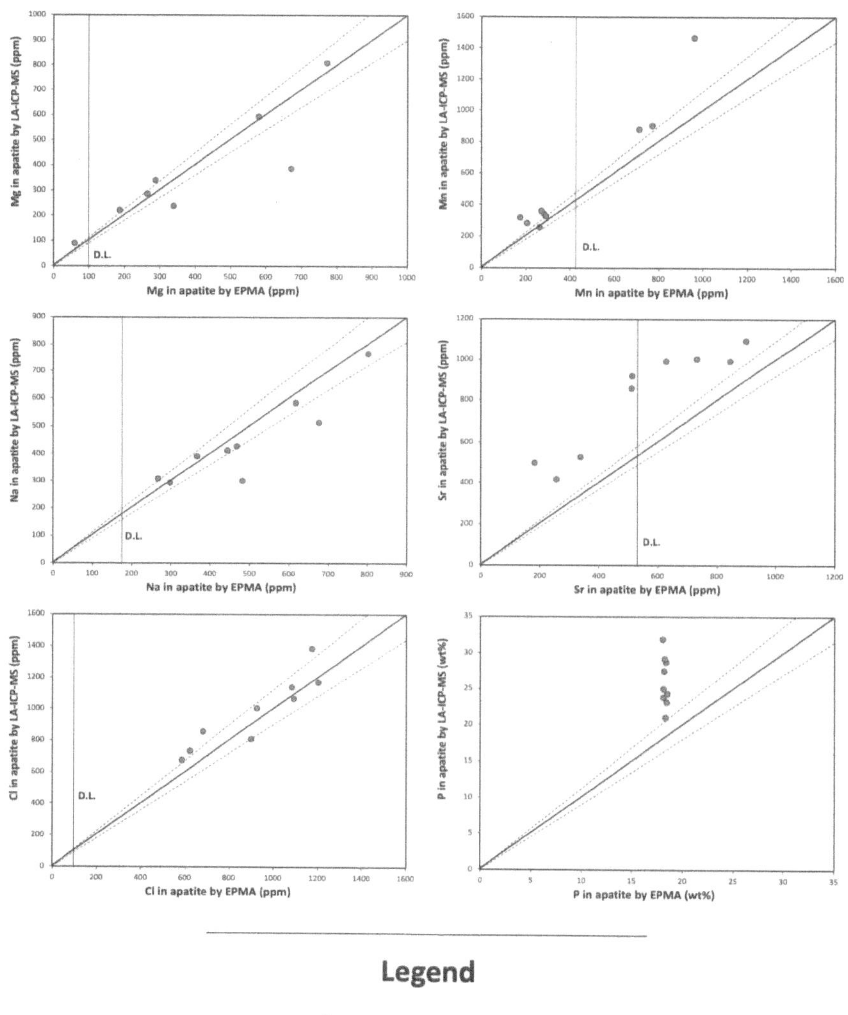

## Legend

● Apatite analysis

**Figure 7.1:** Comparison of average concentrations of an element within apatite by LA-ICP-MS (y axis) and EMPA (x axis). Thick solid line represents a 1:1 slope. Dotted lines represent 10% differences from the 1:1 slope. Solid vertical line represents detection limit of an element by EMPA.

7.2.2   Rare earth element compositions

Apatite chondrite-normalized REE patterns generally show similar enrichments in LREE (light rare earth elements), depletions in HREE (heavy rare earth elements), and moderate to shallow negative Eu anomalies for all four Fe-Ti-P deposits/occurrences studied (Fig. 7.2). However, there is some subtle variation among the deposits. Since there is an abundance of data points from Lac à Paul ore zones relative to the other deposits (n = 428 grains from n = 36 samples), determined by Chartier-Montreuil (2017), REE patterns in apatite from other deposits will be compared with the range of REE abundances in apatite from Lac à Paul. One difference is that apatite from the Lac à Paul ore zones lack a slightly positive Ce anomaly compared to all the other deposits (Fig. 7.2A) which could be attributed to analytical error but, analyses of certified reference materials and in-house standards produced accurate and precise measurements for all REE (Tables 7.3 and 7.4). Chondrite-normalized patterns from St. Charles de Bourget (Fig. 7.2B) and Lac Perron mineralization (Fig. 7.2C) show slightly shallower slopes relative to the Lac à Paul field indicating lower degrees of HREE depletion. However, apatite from St Charles de Bourget (Fig. 7.2B) appears to have a higher relative abundance of REE compared to Lac Perron (Fig. 7.2C). Chondrite normalized apatite patterns from Lac à l'Orignal (Fig. 7.2D) show slightly steeper slopes with similar HREE depletions, but more enriched in LREE when compared to Lac à Paul ore zone. Apatite from Lac à l'Orignal also displays a shallower negative Eu anomaly compared to apatite from the other deposits/occurrences examined in this study (Fig. 7.2D). There is significant variation in total REE within the Lac à l'Orignal deposit; sample MG-LO-01 appears to be the most enriched in REEs while sample LacOrignal is the most depleted (Appendix 6). The stratigraphic location of sample LacOrignal is unknown but, MG-LO-01 represents the stratigraphically highest nelsonitic norite.

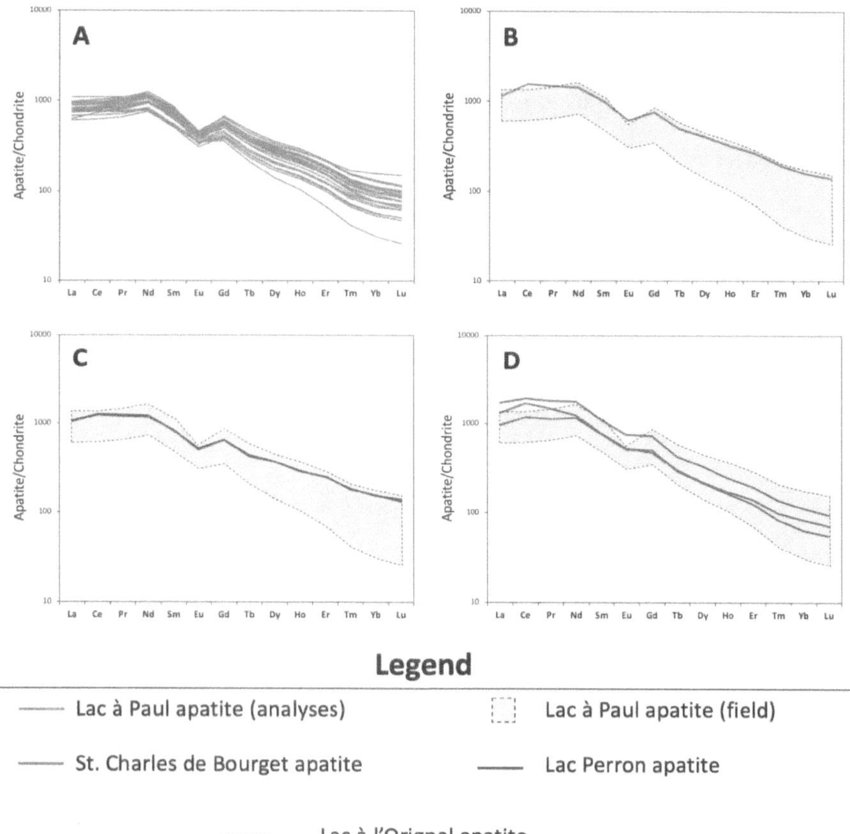



## Legend



**Figure 7.2:** REE concentrations of apatite, normalized to chondrite (Sun and McDonough 1995), from Fe-Ti-P mineralization in Lac St. Jean area. Each line represents the average composition of apatite from a sample (N) normalized to chondrite. Results displayed in figures B-D are compared with range of chondrite normalized concentrations of apatite from the ore zones of the Lac à Paul deposit. A) Lac à Paul (N = 36; data supplemented by Chartier-Montreuil 2017) B) St. Charles de Bourget (N = 3; data supplemented by Desormiers 2015), C) Lac Perron (N = 3) and D) Lac à l'Orignal (N = 3).

To better understand the subtle differences in chondrite-normalized patterns for apatite, REE ratios from all the deposits/occurrences are compared using the chondrite-normalized ratios of LREE to HREE ($La_N/Yb_N$, Fig. 7.3A), LREE to MREE (medium rare earth elements) ($La_N/Sm_N$, Fig. 7.3B), and MREE to HREE ($Gd_N/Yb_N$, Fig. 7.3C) versus the total REE contents of apatite. Figure 7.3 shows that total REE contents in apatite is significantly lower within samples from both ore lithologies (ultramafic nelsonite and nelsonitic norite) at Lac à Paul relative to apatite from the other mineralization examined in this study. For nelsonite, apatite from St. Charles de Bourget is more enriched in total REE than apatite from Lac Perron. Apatite from the nelsonitic norite of Lac à l'Orignal is the most enriched in total REE.

Comparing $La_N/Yb_N$ ratios among the deposits/occurrences indicates that apatites from Lac à l'Orignal, St. Charles de Bourget and Lac Perron display a relatively constant ratio relative to apatite from Lac à Paul ore zones. Analyses from Lac à Paul are much more abundant and were collected over are larger stratigraphic area relative to those from the other deposits. Therefore, the lack in variation in $La_N/Yb_N$ within apatite from the other deposits is either related to primary magmatic conditions or simply not present due to 1) lack of sampling or 2) sampling within a relatively small stratigraphic area. At Lac à l'Orignal, the $La_N/Yb_N$ ratio of apatite remains constant throughout the entire stratigraphy of the deposit but the total REE content within apatite changes. Meanwhile at Lac à Paul (ore zones), there is an increase in the $La/Yb_N$ ratio from ultramafic nelsonite towards nelsonitic norite. This is possibly due to an overall LREE enrichment or HREE depletion during fractionation at Lac à Paul.

There is very little change in $La_N/Sm_N$ ratios within apatite from all the deposits/occurrences indicating that LREE enrichments are likely not the cause of the variable

La$_N$/Yb$_N$ ratios (Fig. 7.2). The nelsonitic norites generally do have larger La$_N$/Sm$_N$ ratios than the nelsonitic peridotites at Lac à Paul but, the variation is insignificant since most apatite analyses plot between La$_N$/Sm$_N$ ratios of 1-2 (Fig. 7.3B).

In contrast, the Gd$_N$/Yb$_N$ ratios within apatite display variation among the deposits/occurrences related to lithology. Fig. 7.3C shows that the lowest Gd$_N$/Yb$_N$ ratios (i.e. flatter slopes) are present within apatite from nelsonites (St. Charles de Bourget and Lac Perron) and the most elevated Gd$_N$/Yb$_N$ ratios (i.e. steeper slopes) are present in apatite from ultramafic nelsonites (Lac à Paul) and nelsonitic norites (Lac à Paul and Lac à l'Orignal). Within the nelsonites, the Gd$_N$/Yb$_N$ ratio of apatite remains relatively constant. In contrast, the Gd$_N$/Yb$_N$ ratio of apatite can vary slightly within Lac à l'Orignal and varies significantly within both ore lithologies of Lac à Paul relative to the other deposits. At Lac à Paul, apatite within the nelsonitic norites generally contain elevated Gd$_N$/Yb$_N$ ratios relative to apatite from the ultramafic nelsonites. From these observations, it can be noted that apatite from ultramafic nelsonites and nelsonitic norites are generally depleted in HREE relative to apatite from nelsonites based on elevated Gd$_N$/Yb$_N$ ratios within the former and depleted Gd$_N$/Yb$_N$ within the latter.

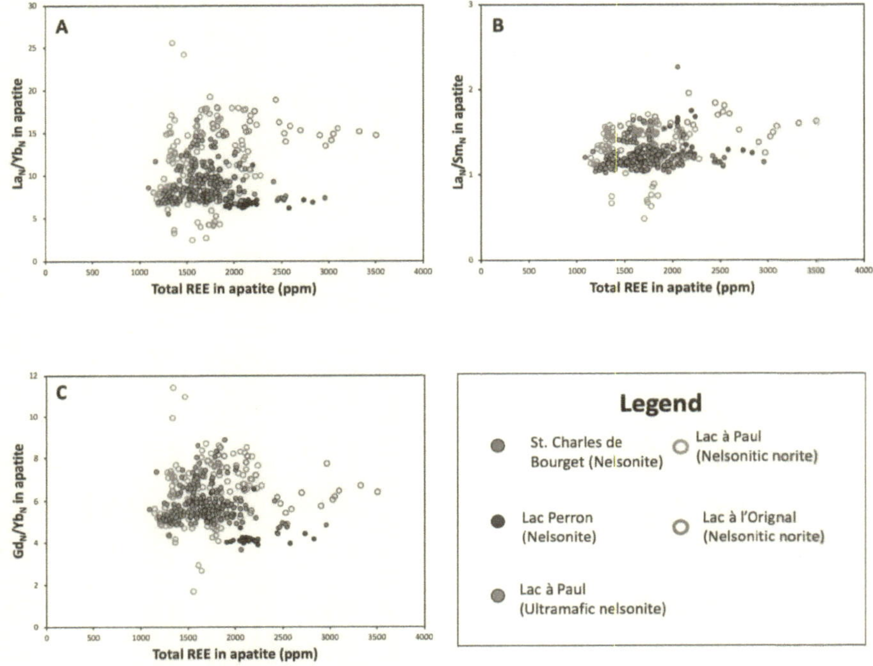


**Figure 7.3:** Binary plots displaying REE element ratios (normalized to chondrite) plotted against total REE concentrations in apatite obtained by LA-ICP-MS. Each point represents an individual apatite analysis. Apatite data from Lac à Paul is supplemented by Chartier-Montreuil (2017) and data from St. Charles de Bourget is supplemented by Desormiers (2015). A) $La_N/Yb_N$ vs. total REE, B) $La_N/Sm_N$ vs. total REE and C) $Gd_N/Yb_N$ vs. total REE.

### 7.2.3   Other trace element abundances in apatite

In this section, the possible connections between concentrations of REE will be compared with that of other trace elements within apatite. Total REE and Sr concentrations within apatite display a slight positive correlation (Fig. 7.4A) among the Fe-Ti-P deposits/occurrences whereas correlations between other trace elements and total REE are absent. Figure 7.4A and C also shows that apatite from Lac à l'Orignal contains the highest Sr contents, followed by apatite from St. Charles de Bourget, the apatite from Lac Perron has the lowest Sr contents. The large dataset of apatite from Lac à Paul displays a large range of Sr contents (237-1314ppm; Figs 7.4 A and C). However, concentrations of Mn (Fig. 7.4B and 7.4C) and Na (Fig. 7.4D) appear to be elevated within apatite from nelsonite samples (St. Charles de Bourget and Lac Perron) relative to apatite from the ultramafic nelsonite (Lac à Paul) and nelsonitic norite (Lac à Paul and Lac à l'Orignal) samples. Although Mn contents of apatite within nelsonite samples from both St. Charles de Bourget and Lac Perron are relatively similar (Fig. 7.4B and C), Na is most elevated within apatite from St. Charles de Bourget nelsonite relative to Lac Perron nelsonite. Concentrations of Mn (Fig. 7.4B and C) and Na (Fig. 7.4D) within apatite remain relatively constant compared to other elements plotted in Figure 7.4 at Lac à Paul and Lac à l'Orignal. These results indicate a sampling size control for Sr and a strong lithological control for Mn and Na contents within apatite. This is likely behind the lack of correlation between Sr and Mn (Fig. 7.4C).

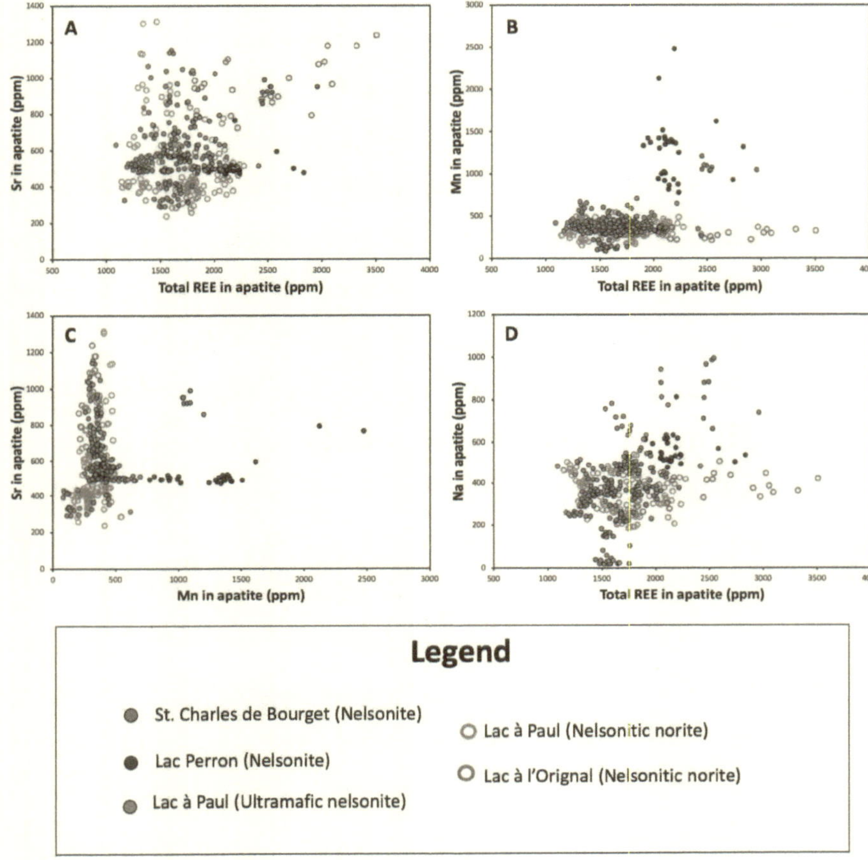

**Figure 7.4:** Binary plots displaying trace element concentrations of apatite. Each point represents an individual apatite analysis. Data for Lac à Paul supplemented by Chartier-Montreuil (2017) and data for St. Charles de Bourget supplemented by Desormiers (2015). A) total REE vs Sr, B) Total REE vs Mn, C) Sr vs Mn, D) Total REE vs Na.

Since the concentration of Cl within apatite is accurate by LA-ICP-MS (Fig. 7.1), possible correlations between volatile and trace element concentrations within apatite are examined in Fig. 7.5. Apatite from Lac à l'Orignal is generally the most enriched whereas apatite from Lac à Paul is generally the most depleted in Cl (Fig. 7.5). The Cl concentrations of apatite from nelsonites remain relatively constant when compared to Cl concentrations within apatite from ultramafic nelsonites and nelsonitic norites (Fig. 7.5). Fig. 7.5 also shows that neither Na (Fig. 7.5A) or Mn (Fig. 7.5B) contents of apatite display any correlation with Cl contents of apatite. However, Sr (Fig. 7.5C) and total REE (Fig. 7.5D) display a dispersed positive correlation with Cl in apatite even on the deposit scale. This indicates that the controls on Cl, Sr, and REE concentrations of apatite are relatively similar.

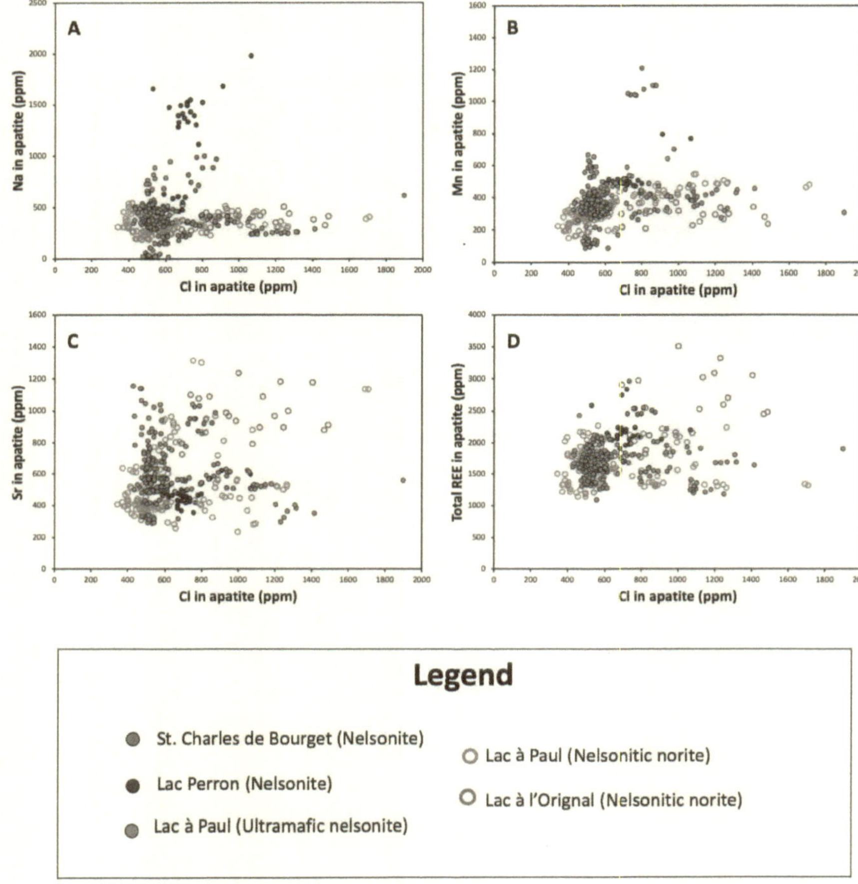


**Figure 7.5**: Binary plots displaying the concentrations of trace elements within apatite plotted against Cl. Each point represents an individual apatite analysis. Data for Lac à Paul from Chartier-Montreuil (2017) and data for St. Charles de Bourget supplemented by Desormiers (2015). A) Na vs. Cl, B) Mn vs. Cl, C) Sr vs. Cl, D) Total REE vs Cl.

136

# Chapter 8  Whole Rock Analysis

Whole rock analysis (WRA) was performed on a total of 45 samples representing the different lithologies from both the mineralization (n =31) and host rocks (n =14), where possible, from each of the 6 deposits/showings. Mineralized samples comprise massive oxide (n=11) from all deposits except Lac à Paul massive nelsonite from St. Charles de Bourget and Lac Perron (n=7), Fe-oxide-rich norite from Lac Margane/Houlière-Nord (n=6), nelsonitic norites from Lac à Paul and Lac à l'Orignal (n=4) and semi-nelsonites (n=3) from Lac à Paul. These semi-nelsonites from the New Zone of Lac à Paul include dunitic nelsonite (n=1), nelsonitic olivine norite (n=1), and nelsonitic pyroxenite (n=1). Associated host rock samples include anorthosite (n=8), from all deposits/occurrences except Lac à Paul and Lac Margane/Houlière, norite (n=3) from Lac Margane/Houlière, and ferrodiorite dykes (n=3) from St. Charles de Bourget and Lac à Paul. Results for whole rock analysis are displayed in Appendix 7.

In this section, the analytical methods, quality of the data collected, and geochemical trends of major and minor elements related to fractionation and mineralogy within the samples are described for each lithology. Geochemical trends from mineralized samples by WRA further compliments LA-ICP-MS data by determining effects of 1) modal proportion of mineralization (Fe-oxides ± apatite) to primary silicate minerals (plagioclase, olivine, orthopyroxene) when present, 2) Fe-oxide speciation (relative modal proportions of magnetite, ilmenite, and spinel), and 3) presence and modal proportion of apatite on REE contents. The geochemical trends for unmineralized host rocks (anorthosites, norites, and ferrodiorites) provides additional constraints on 1) fractionation of host rocks on the regional scale and 2) parental melt compositions of each deposit/occurrence (Chapter 8). Unfortunately, ferrodiorite samples from only 2 locations (St. Charles de Bourget in the South and Lac à Paul in the North) could be sampled and analyzed by WRA.

## 8.1  Analytical Methods

First, the samples were crushed into a fine powder using a steel jaw crusher followed by a ceramic disk mill at the University of Ottawa. After each sample was crushed, cleaning was performed by pulverizing pure silica sand with the ceramic disk mill at least two times. To test for contamination, after select samples, silica sand (n=4) was then pulverized after cleaning of the ceramic disk mill. The resulting silica powder would be sent for whole rock analysis with the powdered samples. Analyses of silica blanks indicated that contamination occurring during this process is negligible.

One gram of each pulverized sample or silica blank was weighed out and sent to ALS Canada Ltd. for whole rock analysis. Standards analyzed with the unknowns include KPT-1 (a diorite dyke from the Sudbury Igneous Complex: Webbe et al. 2006). Whole rock analysis was carried out using three methods of analysis were used to obtain the results used in this project; 1) Loss on ignition (LOI), 2) lithium metaborate fusion followed by XRF (X-Ray Fluoresence :analysis code ME-XRF21u) which is optimal for major and minor element oxide concentrations and 3) four acid 'near total' digestion followed by ICP-MS (analysis code MER-MS61r) which is optimal for low abundance trace elements (such as REE and HFSE). Details of the analytical procedures are outlined on the ALS Canada Ltd. website and summarized here. Loss on ignition was obtained using a thermogravimetric analyzer within a furnace. The method for the major element package involved mixing 0.2 g of sample along with 0.9 g of a lithium metaborate flux which was heated at 1000°C to fuse the mixture together and form a glass disk which is then analyzed by XRF. The method for the trace element package was performed by dissolving the crushed powder in a mixture of 4 acids

(HNO$_3$, HCl, HF, and H$_2$SO$_4$). The dissolution is 'near total' but does not completely digest some resistant accessory minerals (barite, zircon, some REE-oxides, columbite-tantalite, Ti, Sn, and W). The resulting solution was then analyzed by ICP-MS. Results for standards and silica blanks are located in Appendix 7.

8.2   Results

For graphing purposes, analyses that registered concentrations below detection limit were converted to ½ the detection limit for that respective element. As there are issues that exist regarding the digestion of certain resistant elements before analysis by ICP-MS, it is important to evaluate the effect of these issues before interpreting the geochemical trends of these samples.

8.2.1   Comparison of analysis methods

Before plotting the whole rock data, we first must consider which analysis method provides the most accurate and precise results for each element as we have many concentrations of elements obtained by both analytical methods. For major elements and some trace elements (Cr, V) within these deposits, the results obtained from XRF (complete digestion) are compared with the results obtained by ICP-MS (near-complete) to see if they agree with each other to within 10% (optimal conditions). Results obtained from XRF and ICP-MS for KPT-1 for all elements appear to agree within 10% (Appendix 7). However, to better evaluate whether 4-acid digestion is suitable for the analysis of massive oxides, data from BC-28 (massive magnetite from the Bushveld Complex, with minor (5% ilmenite and no spinel exsolutions) must be examined. Analysis of BC-28 shows that concentrations of most elements obtained by both methods agree within 10%, however, Al, Mg, and Zn do not which possibly indicates incomplete dissolution of sub-microscopic spinel phases.

Comparison of element concentrations in all samples determined by the two methods are displayed in Fig. 8.1 and Appendix 7. Elements that agree within 10% by both analyses are Ca, Fe, K, Sr and V. However, Al by both methods does not agree within 10% only in anorthosites, massive oxides, and nelsonites and Mg does not agree within 10% only in massive oxides and nelsonites. Elements that can sometimes display lower concentrations by ICP-MS are Cr, Co, Mn, Ni, and Ti and elements which do not agree by both methods are Zn and Zr.

In most cases besides Fe (Fig. 8.1A), analyses by ICP-MS can display consistently lower values than those by XRF. This is likely due to incomplete dissolution of ilmenite and spinel by the 4-acid mixture before ICP-MS takes place. These effects are most pronounced in Ti (Fig. 8.1B), V (Fig. 8.1C), and Cr (Fig. 8.1D) as results deviate outside the 10% field when plotting near the upper detection limit for Ti and >1000 ppm for Cr and V. Although for Ti, most massive oxide samples could not be plotted because concentrations obtained by ICP-MS are regularly above the upper detection limit. For Al, the only time that the results do not plot within the 10% field are within the range of 2-5% and >14%, likely within the massive samples (oxides/nelsonites) for the former and anorthosites for the latter (Fig. 8.1E). This is likely due to possible incomplete dissolution of spinel within the massive samples and plagioclase within the anorthosites. The incomplete dissolution of spinel could also affect the Mg results as values within 1-3 % Mg consistently plot outside the 10% field (Fig. 8.1F). Therefore, the values obtained by XRF are preferentially used for all transition metals. For elements with heavy atomic numbers, such as Sn and Pb, we used the ICP-MS values for graphing since the analyses by XRF are too close or below the lower detection limit.

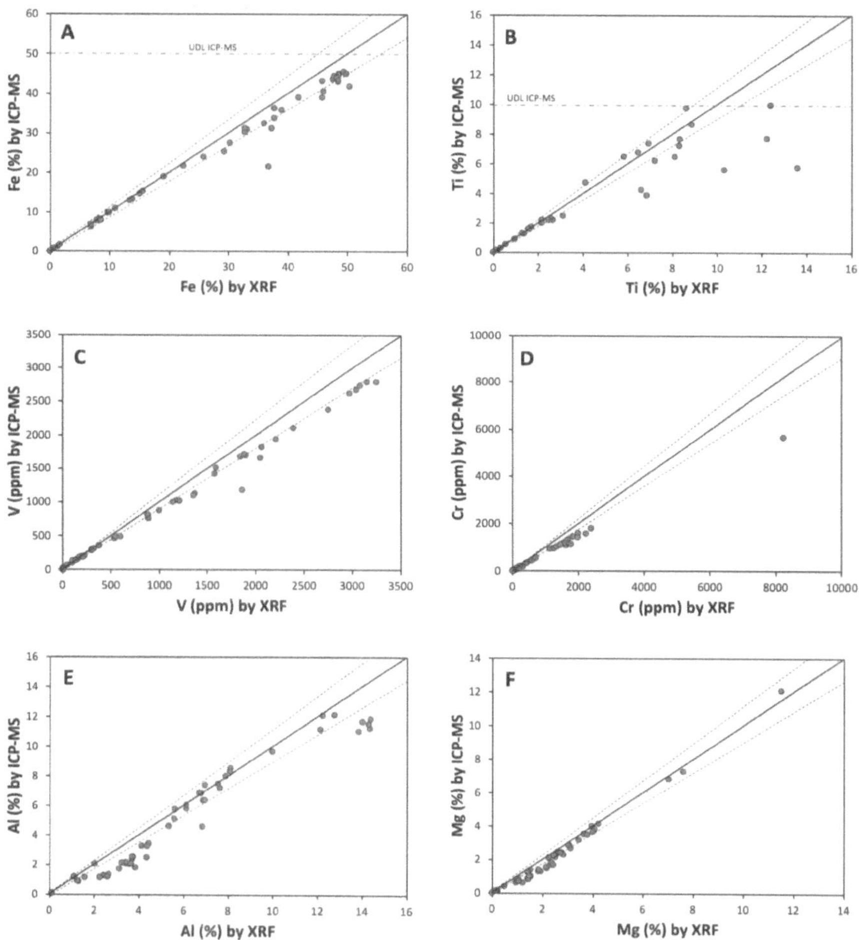

**Figure 8.1:** Binary plots showing the concentrations of elements in whole rock analysis by both analytical methods. Solid black line indicates a 1:1 slope, dotted black lines indicate 10% error from the 1:1 slope and the blue dotted line notes the upper detection limit (UDL) of an element by ICP-MS.

### 8.2.2    Major element compositions

To constrain the major element trends, the results were plotted on Harker diagrams using major element oxides that best constrain compositional trends within these samples (Fig. 8.2). The compositions of major silicate minerals (olivine, orthopyroxene, and plagioclase) were also annotated on each Harker diagram in Figure 8.2 to constrain the effect of silicate minerals on the whole rock composition of each sample. However, most of the changes in major element chemistry are likely due to modal percentages of the ore minerals (Fe-oxides and apatite) relative to primary silicate minerals (plagioclase, olivine, orthopyroxene).

The $SiO_2$ contents for each lithological group (Fig. 8.2) are relatively consistent with massive mineralized samples (oxides and nelsonites) displaying the lowest (<10 wt%), and anorthosites displaying the highest $SiO_2$ contents. Of the major element oxides, $Fe_2O_{3(total)}$ (Fig. 8.2A) and $TiO_2$ (Fig. 8.2B) display a negative correlation with $SiO_2$, CaO (Fig. 8.2C) and $Al_2O_3$ (Fig. 8.2D) display a positive correlation with $SiO_2$, and MgO (Fig. 8.2D) and $P_2O_5$ (Fig. 8.2F) display no significant correlation with $SiO_2$. The pattern displayed on (Figs 8.2A and B) is predominantly due to the decreasing modal percentages of Fe-oxide phases from massive mineralization to anorthosites. However, the overall trend can vary due to the presence of both mafic silicates (olivine, pyroxene) and apatite. The effect of Fe-Mg silicates within semi-massive mineralization, such as nelsonitic norites (Lac à Paul and Lac à l'Orignal) and nelsonitic ultramafic lithologies (Lac à Paul), on the Fe content of these samples is not clear as apatite will dilute the Fe content and Fe-Mg silicates will increase it at the same time. The increasing presence of coexisting silicate phases can also affect the Ti content since Ti in most silicates noted in this study is negligible. Therefore, increasing the modal percentage of silicate minerals while subsequently decreasing Fe-oxide modal percentage should decrease whole rock Ti contents. This means that the overall Ti content within a sample is likely determined by the modal percentage of Ti-magnetite or ilmenite.

The CaO concentration within samples are likely due to the modal percentage of both plagioclase and apatite which can explain the 2 different trends displayed on (Fig. 8.2C). Clinopyroxene was not considered an important control on whole rock Ca as it is present within negligible quantities in this study. Calcic amphibole was also not considered as an important control on whole-rock Ca as it is considered a secondary silicate phase which originated from re-equilibration and/or metamorphism and therefore, its presence should have negligible effects on the whole rock Ca of the protolith. The main trend follows a positive correlation starting from low CaO in massive oxides to high CaO in anorthosites and the second trend is a minor negative correlation starting from apatite-rich, silicate-poor nelsonites to apatite-free anorthosites (Fig. 8.2C). The main Al-rich mineral phases in these rocks are plagioclase and the pleonaste component of magnetite (53.57-61.25 $Al_2O_3$ wt.%), hosted now either in exsolutions of Al-spinel and/or in solid solution in magnetite). Knowing this, the positive correlation from massive mineralization (0.13-10.35% $SiO_2$ 4.28-8.22% $Al_2O_3$) to anorthosites (high-Al) is the main trend visible (Fig. 8.2D). Deviation from this trend to lower Al contents is by dilution from the presence of Al-poor silicates such as olivine and orthopyroxene or apatite, as demonstrated by ultramafic nelsonites and pyroxene-rich gabbros.

Apatite, the only P-bearing phase observed during petrography, is only present within nelsonitic samples and thus $P_2O_5$ contents are dependent on the presence or modal proportions of apatite which can be highly variable within samples from Fe-Ti-P deposits/occurrences. As noted in Chapter 4, nelsonite samples from St. Charles de Bourget appear to be considerably enriched in apatite relative to nelsonite from Lac Perron (Fig. 8.2F). However, that could be attributed to the heterogeneous distribution of very coarse-grained apatite within Lac Perron.

The concentration of MgO within these samples is hosted in 2 mineral groups, 1) Fe-Mg silicates and 2) pleonaste component of magnetite (1.59-6.74 wt% MgO). The modal proportions of these minerals are extremely variable within these lithologies as noted by petrography thus explaining the general lack of correlation between MgO and $SiO_2$ contents. The samples plot closer to the tie line of oxide and plagioclase with only minor contribution of Mg from mafic minerals.

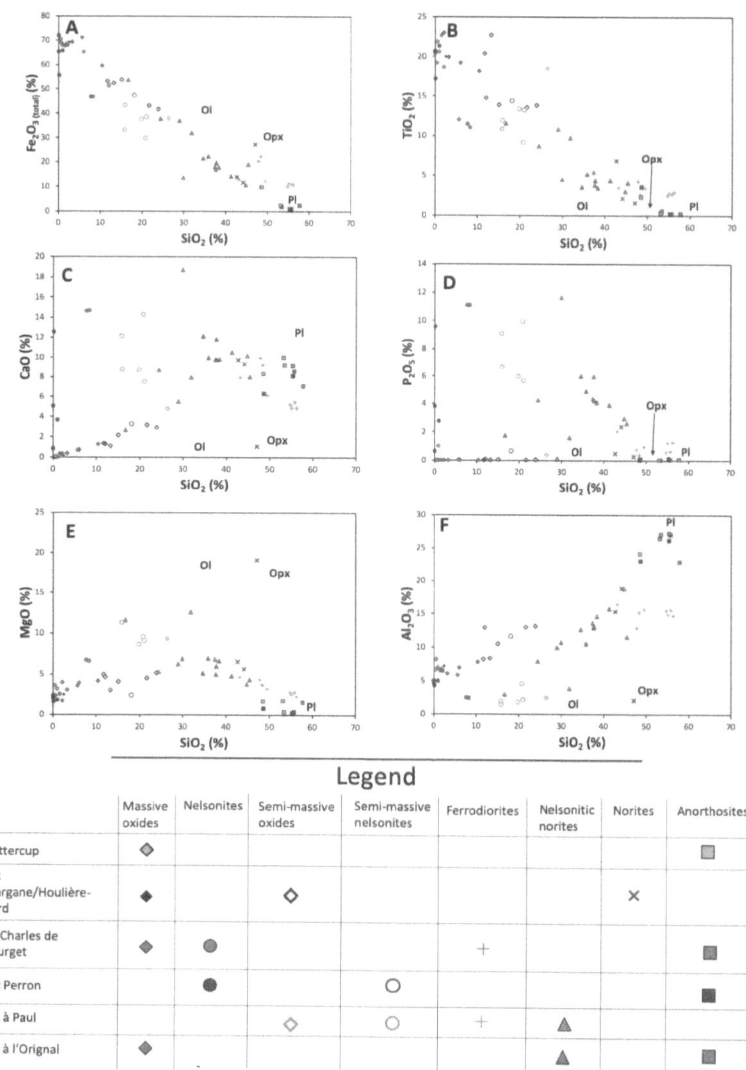



## Legend

| | Massive oxides | Nelsonites | Semi-massive oxides | Semi-massive nelsonites | Ferrodiorites | Nelsonitic norites | Norites | Anorthosites |
|---|---|---|---|---|---|---|---|---|
| Buttercup | ◇ | | | | | | | ▧ |
| Lac Margane/Houlière-nord | ◆ | | ◇ | | | | × | |
| St. Charles de Bourget | ◆ | ● | | | + | | | ▧ |
| Lac Perron | | ● | | ○ | | | | ■ |
| Lac à Paul | | | ◇ | ○ | + | ▲ | | |
| Lac à l'Orignal | ◆ | | | | | ▲ | | ▧ |

**Figure 8.2:** Harker diagrams showing the compositional trends within the mineralization, host, and associated lithologies examined during this project. Grey fields plotted on these diagrams are obtained from EMPA data using the range of compositions obtained by EMPA for the respective silicate. A) Fe₂O₃(total) vs SiO₂ B) TiO₂ vs. SiO₂ C) CaO vs SiO₂ D) Al₂O₃ vs. SiO₂ E) MgO vs. SiO₂ F) P₂O₅ vs. SiO₂. Plagioclase (Pl), olivine (Ol), orthopyroxene (Opx). Data for Lac à Paul P deposit supplemented from Chartier-Montreuil et al. (2017).

8.2.3    V-Ti-P composition of mineralization

To compare the main commodities within these deposits (Ti, V, and P), each of these elements are plotted against the total $Fe_2O_3$ content (i.e., Fe-oxide content) of each sample in Fig. 8.3. Both $TiO_2$ (Fig. 8.3A) and V (Fig. 8.3B) are positively correlated with $Fe_2O_{3(total)}$, as expected because these elements are hosted in Fe-oxides, and that $P_2O_5$ and $Fe_2O_{3(total)}$ display no significant correlation (Fig. 8.3C). As described below, the overall content of Ti and V within the ore samples is dependent on the modal percentages and speciation of Fe-oxide minerals while $P_2O_5$, which is controlled by apatite, is mostly independent on the abundance of Fe-oxides.

The positive correlation of $TiO_2$ with $Fe_2O_{3(total)}$ is evident in Fig. 8.3A. To quantify the effect of Fe-oxide speciation (magnetite vs. ilmenite) within these samples, two slopes are constrained using the following data; 1) end-member titanomagnetite, using the magnetite composition from Buttercup (average of whole rock of massive oxides) and 2) end-member hemo-ilmenite using the whole-rock values from Charlier et al. (2008) of massive hemo-ilmenite from the Lac Tio deposit. These two Fe-oxides form the dominant mineralization in anorthosite complexes (Hébert et al. 2005, Woodruff et al. 2013, Charlier et al. 2015). Buttercup was chosen because the massive oxides samples typically contained >90% magnetite and contained magnetite with a $TiO_2$ content of ~20 wt.%. The whole rock concentrations for all Fe-oxide-bearing lithologies (all lithologies except for unmineralized norite and anorthosite) hosted by the Lac St Jean anorthosite closely follow the titanomagnetite trend defined by Buttercup (Fig. 8.3A). This is interesting because several of these deposits contain significantly different magnetite/ilmenite ratios to Buttercup (e.g. Lac Margane/Houlière has 50% ilmenite but same whole rock composition as Buttercup with only 5% ilmenite). Massive oxide from Lac à l'Orignal (Fe-Ti-P), which is hosted within the Lac

Vanel anorthosite, significantly plots below the Lac St. Jean trend due to the low-Ti content of titanomagnetite (12.05 wt.% $TiO_2$: Fig. 8.3A). Although hemo-ilmenite is present within nelsonitic norites from Lac à l'Orignal, these samples do not plot on the Lac Tio hemo-ilmenite trend on (Fig. 8.3A). This is possibly due to low modal percentages of hemo-ilmenite (5-10%) relative to plagioclase and orthopyroxene. However, increasing modal percentages of plagioclase and apatite within samples should have similar effects on Fe and Ti whole rock contents as both Fe and Ti will be diluted. Furthermore, increasing modal percentages of Fe-Mg silicates can dilute whole rock Ti contents within a sample as Fe-Mg silicates contain negligible Ti contents. This was not corrected for in the final graphing value because the effects are believed to be minor and only noted within a minority of the samples.

Although V and $Fe_2O_{3(total)}$ are positively correlated, oxide-rich samples from Fe-Ti-V and Fe-Ti-P deposits/occurrences display 3 separate trends (Fig. 8.3B). The steeper, V-rich trend is defined by massive ores from Fe-Ti-V mineralization (i.e., Buttercup and Lac Margane), an intermediate V concentration trend is projected from massive oxides and nelsonites from Fe-Ti-P mineralization (St. Charles de Bourget and Lac à l'Orignal), and a V-poor trend is projected from massive nelsonite ore from Lac Perron Fe-Ti-P mineralization (Fig. 8.3B). As expected, massive samples from Fe-Ti-V mineralization tend to be richer in V than the Fe-Ti-P mineralization due to fractional crystallization. However, this is also likely to be attributed to dilution since apatite does not contain any V or Fe.

The main control on the $P_2O_5$ content within a sample is the modal proportions of apatite which is either cumulus (in P-rich samples) or intercumulus, having crystallised from trapped liquid in P poor samples. The amount of trapped liquid seems to be negligible in samples which do not contain visible apatite, as most of these samples contain near zero concentrations of $P_2O_5$ (Fig. 8.3C). Concentrations of $P_2O_5$ contents are the highest within nelsonites from St. Charles de Bourget (11-11.45 % $P_2O_5$) and nelsonitic ultramafic units from Lac à Paul (5.65-11.65 % $P_2O_5$) and nelsonitic norite from Lac Orignal (4.22-5.98 % $P_2O_5$: Fig. 8.3C). The wide variation of $P_2O_5$ contents within massive nelsonite from Lac Perron is indicative of the heterogeneous distribution of very coarse-grained apatite within the ore body, which was noted during field observations (section 4.2.2).

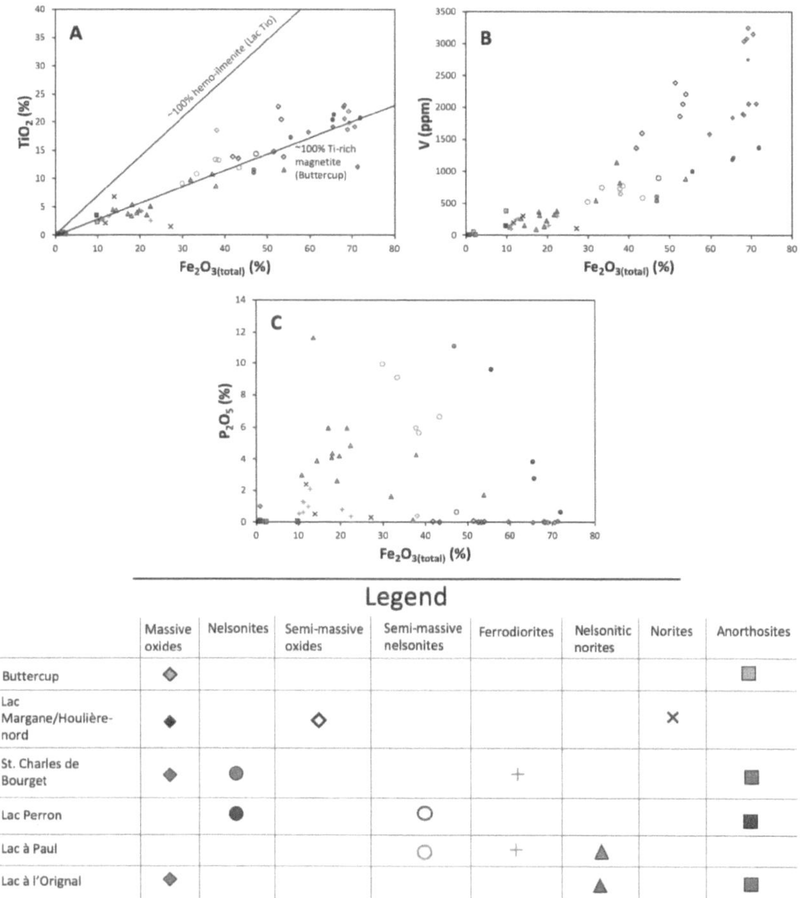

## Legend

| | Massive oxides | Nelsonites | Semi-massive oxides | Semi-massive nelsonites | Ferrodiorites | Nelsonitic norites | Norites | Anorthosites |
|---|---|---|---|---|---|---|---|---|
| Buttercup | ◇ | | | | | | | ▦ |
| Lac Margane/Houlière-nord | ◆ | | ◇ | | | | × | |
| St. Charles de Bourget | ◆ | ● | | | + | | | ▦ |
| Lac Perron | | ● | | ○ | | | | ▦ |
| Lac à Paul | | | | ○ | + | ▲ | | |
| Lac à l'Orignal | ◆ | | | | | ▲ | | ▦ |

**Figure 8.3:** Binary plots displaying the mineralization content of samples analyzed during this project. Additional data for Lac à Paul Fe-Ti-P deposit is from Chartier-Montreuil et al. (2017.). A) $TiO_2$ vs $Fe_2O_{3(total)}$, the 100% magnetite trend is $TiO_2$ and $Fe_2O_3$ contents of magnetite obtained from LA-ICP-MS from sample BCP-2 (~100% magnetite). Hemo-ilmenite trend defined by whole-rock values of ore from the Lac Tio Ti deposit, (Havre-St. Pierre anorthosite suite, Québec; Charlier et al. 2008). B) V vs. $Fe_2O_{3(total)}$ C) $P_2O_5$ vs. $Fe_2O_{3(total)}$.

149

8.2.4   Rare earth elements within nelsonitic samples

The whole rock REE patterns of nelsonites from St. Charles de Bourget and Lac Perron show similar HREE depleted slopes with negative Eu anomalies (Fig. 8.4A). However, REE are more abundant within nelsonite from St. Charles de Bourget relative to nelsonite from Lac Perron. Furthermore, REE abundances are similar between the nelsonite samples at St. Charles de Bourget but vary significantly between nelsonites at Lac Perron. This is likely due to the abundance and distribution of apatite among the nelsonite samples. Nelsonites from St. Charles de Bourget significantly contain a higher modal percentage of apatite (45%) relative to nelsonites from Lac Perron (25-45%). Furthermore, fine-grained apatite within St. Charles de Bourget nelsonite outcrops display an even distribution whereas apatite within nelsonite outcrops at Lac Perron occurs in local, very coarse-grained clusters (Fig. 4.9B). This indicates that apatite from Lac Perron nelsonites is heterogeneously distributed relative to apatite from St. Charles de Bourget nelsonites and can explain the differences in REE abundances within the Lac Perron nelsonite samples. This is displayed on Fig. 8.4B as REE and $P_2O_5$ are positively correlated and nelsonites from St. Charles de Bourget are more enriched in both total REE and $P_2O_5$ relative to nelsonites from Lac Perron.

The REE distribution within apatite-bearing samples from Lac à Paul (New Zone) and Lac à l'Orignal is more difficult to explain since these samples contain silicate phases (plagioclase, orthopyroxene, amphibole) which could alter the REE signature displayed by apatite. This affect appears to be negligible within the nelsonitic norites from Lac à l'Orignal as they display chondrite-normalized REE patterns similar to the apatite shown in Fig. 7.2 with enrichments in LREEs, shallow negative Eu anomalies and strong HREE depletions (Fig. 8.5A). The chondrite-normalized patterns of samples from Lac à Paul appear different from the others plotted on Fig. 8.4A as they display shallower HREE depletions. If the presence of silicates, such as plagioclase and orthopyroxene, have little effect on samples from Lac à l'Orignal, it is likely they also have little effect on samples from Lac à Paul. However, amphibole present within low grade mineralization (nelsonitic norites) at the Lac à Paul deposit was no was noted to host 30% of the total HREE budget based on mass balance calculations by (Chartier-Montreuil et al. 2017). Therefore, shallow HREE depletions within Lac à Paul apatite-bearing samples indicate an addition of HREE to the whole rock analysis and are possibly due to 1) lower modal percentage of apatite relative to Lac à l'Orignal (Fig. 8.4B), which is LREE enriched (Figs. 7.2 and 7.3) or 2) possible presence of HREE-bearing amphibole or accessory phases (e.g., zircon) undetected by petrography.

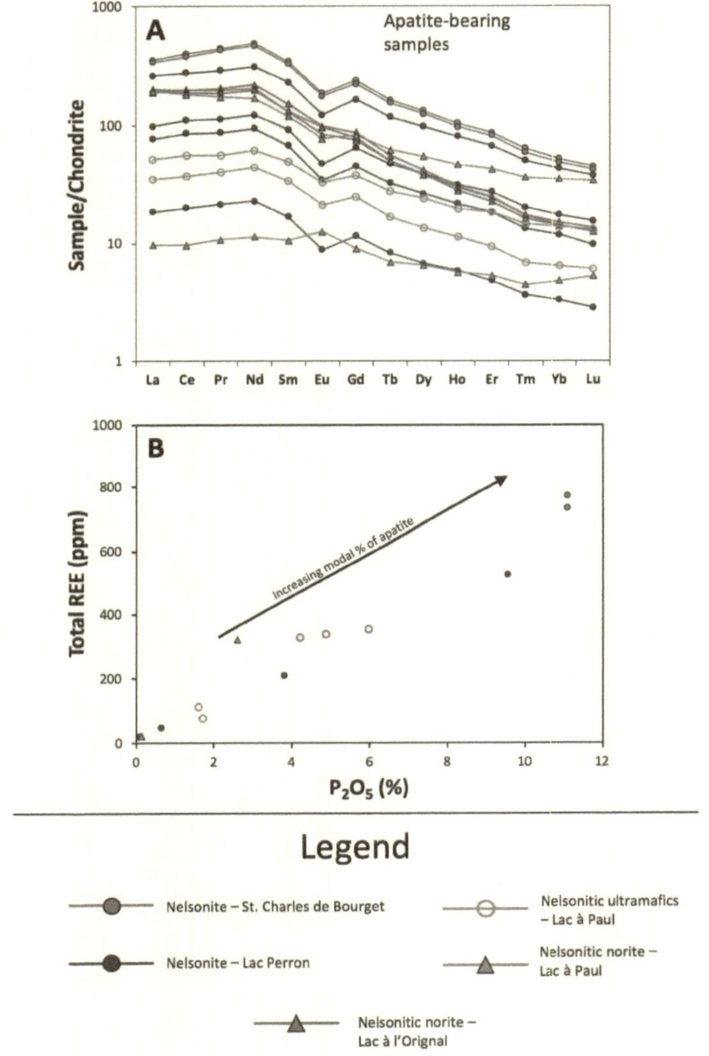


**Figure 8.4:** Geochemical plots displaying whole rock compositions of apatite-bearing samples. A) Chondrite-normalized plots, normalization values obtained from Sun and McDonough (1995). B) Total REE vs $P_2O_5$.

152

8.2.5    Rare earth element patterns of host rocks and ferrodiorites

The REE contents, normalized to chondrite, were examined for all anorthosite (host rock) samples (Fig. 8.5A), ferrodiorite (Fig. 8.5B), and unmineralized norite samples (Fig 8.5C). All anorthosites display a similar pattern, as they are LREE-enriched, HREE-depleted and display a strong, positive Eu anomaly (Fig. 8.5A), typical of plagioclase. Overall, total REE contents are generally higher in anorthosites hosting Fe-Ti-P mineralization relative to anorthosites hosting Fe-Ti-V mineralization. This is shown by anorthosites from Buttercup displaying the lowest REE abundances relative to those from St. Charles de Bourget, Lac Perron, and Lac à l'Orignal (Fig. 8.5C). Among the Fe-Ti-P deposits/occurrences, anorthosite from Lac à l'Orignal is the most REE-enriched, followed by Lac Perron, then St. Charles de Bourget (Fig. 8.5A). The HREEs are also significantly more depleted in anorthosites from St. Charles de Bourget and Buttercup relative to the others (Fig. 8.5A). However, the HREE pattern for all anorthosites in relatively flat except for one anorthosite from near the Buttercup deposit (sample MG-BCP-08). The erratic HREE pattern shown by sample MG-BCP-08 can potentially be explained by the below detection limit concentrations of some HREE within this sample

Chondrite-normalized trends of ferrodiorites display enrichments in LREE with depleted HREE and shallow negative to flat Eu anomalies (Fig. 8.5B). Depletions in HREE are significantly stronger at Lac à Paul relative to St. Charles de Bourget. Furthermore, HREE depletions within ferrodiorites at St. Charles de Bourget are similar whereas, HREE depletions within Lac à Paul ferrodiorites display significant variation (Fig. 8.5B). This indicates that HREE appear to fractionate at Lac à Paul but not at St. Charles de Bourget.

Unmineralized gabbros/norites from Lac Margane/Houlière-Nord are significantly more HREE-enriched closer to the mineralization (Fig 8.5C). Sample MG-LMG-03b (near mineralization), displays a relatively flat chondrite-normalized pattern with enrichments in MREE whereas MG-LMG-04 (south of mineralization) is relatively enriched in LREE and depleted in HREE (Fig. 8.5C). Within norite units away from mineralization at Lac Margane/Houlière-Nord, relict pyroxenes are present. However, closer to mineralization, pyroxenes have been entirely converted to hornblende. This could possibly result in the different chondrite-normalized patterns shown on Figure 8.5C.

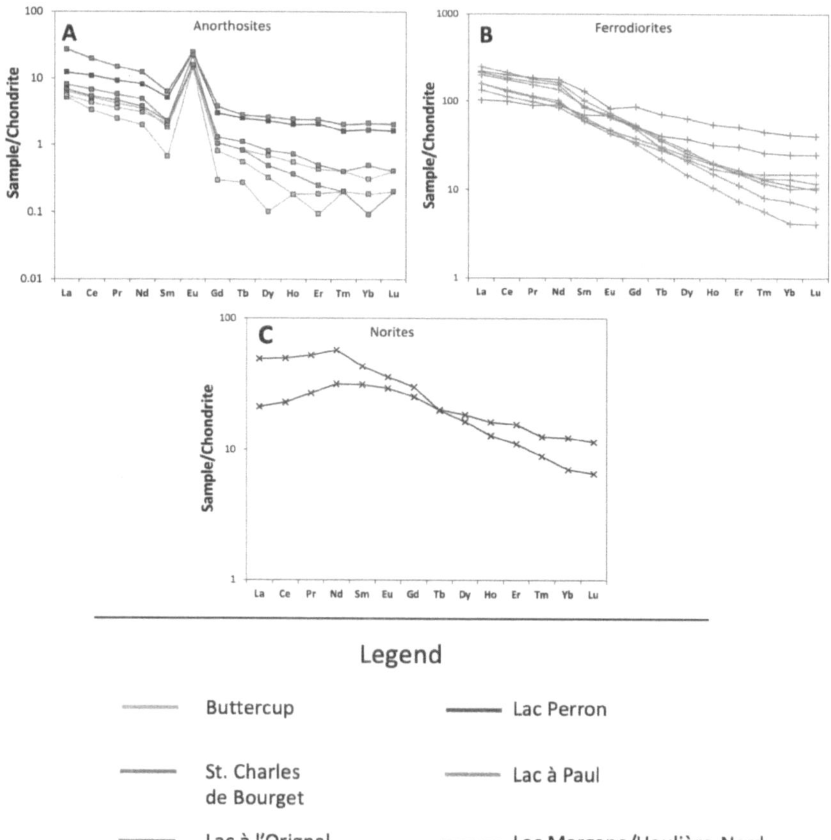



**Figure 8.5:** Chondrite-normallized plots for host rocks (A and C) and associated lithologies (B). A) Host anorthosite near (within 1-2km) the mineralized zones. B) Ferrodiorites found near the mineralized zones. C) Norites from Lac Margane/Houlière/Nord. Normalization values are from Sun and McDonough (1995). Data for Lac à Paul ferrodiorites supplemented from Chartier-Montreuil (2017).

# Chapter 9  Discussion

Magnetite and apatite have commonly been used as petrogenetic indicators within magmatic oxide deposits hosted by 1) layered intrusions, e.g. Bushveld Complex, South Africa (Barnes et al 2004); Sept Iles, Canada (Namur et al. 2010); Panzhihua, China (Pang et al. 2007) and 2) massif anorthosites, e.g. Tellnes, Norway (Charlier et al. 2006); Suwalki, Poland (Charlier et al. 2009); Grader Lake, Canada (Charlier et al. 2008); and Damiao, China (He et al. 2016). Compared to layered intrusions, anorthosite complexes and their oxide mineralization generally: 1) Lack clear stratigraphic relations between oxide mineralization and host mafic rocks; 2) Underwent high-grade metamorphism (amphibolite-granulite facies) and textural coarsening (Higgins 2011); and 3) Contain abundant coarse-grained Al-spinel (>1 cm size) in host Fe-oxide, which is not common within magmatic oxide deposits within layered intrusions where Al-spinel is sub-mm in size.

This project aims to first, evaluate if magnetite and apatite can still be used as petrogenetic indicators of magmatic processes (magma composition, fractional crystallization, $fO_2$ conditions etc.) for oxide mineralization in anorthosite complexes hosted in granulite terrains, such as the Lac St. Jean anorthosite complex, whereby post-magmatic processes (diffusive modification) were probably enhanced during metamorphism. The second aim of the project is to use magnetite and apatite chemistry to better understand the relationships between the 5 oxide deposits/occurrences studied within the Lac St. Jean anorthosite suite based on spatial association, magma composition, degree of fractionation, and redox ($fO_2$) conditions. Noting that Fe-oxide deposits/occurrences hosted within the (1165-1135 Ma) Lac St. Jean anorthosite are typically magnetite-dominated, with variable amounts of ilmenite and apatite, whereas the one mineralized location studied (Lac à l'Orignal Fe-Ti-P) from the younger (1080 ± 2 Ma; van Breeman 2009) Lac Vanel anorthosite typically contains an Fe-oxide assemblage of hemo-ilmenite and magnetite, it is also important to try to constrain the underlying cause for the differences in mineralogy (e.g., magma composition, redox conditions). In this section, the potential of using in-situ chemistry of Fe-oxide minerals and apatite, combined with classic tools such as field observation, petrography, and whole-rock geochemistry, to provide constraints on the genesis of magmatic Fe-Ti-V-P mineralization within the Lac St. Jean area will be examined.

## 9.1 Evaluating magnetite chemistry as a petrogenetic indicator for anorthosite-hosted oxide deposits

### 9.1.1 Previous work/background

The primary (magmatic) composition of igneous magnetite is controlled by 1) the partition coefficient ($D_{Melt}^{Mt}$) of an element into magnetite; 2) the concentration of an element in the parental magma, which changes due to fractional crystallization; and 3) co-crystallizing minerals and paragenetic sequence, all of which is dependent on temperature, pressure, fO$_2$, and volatile content of the melt (Snyder et al., 1993; Toplis and Carroll 1995; Toplis and Corgne 2002; Botcharnikov et al 2008; Dare et al., 2012, 2014). However, the primary magmatic composition can be modified, after crystallization, by sub-solidus processes (diffusive modification with other Fe-oxides or Fe-Mg silicates) both during cooling of the intrusion and later by metamorphism.

Previous work on un-metamorphosed layered intrusions, such as the Upper Zone of the Bushveld Complex and the Layered Series of Sept Iles intrusion, has shown that during fractional crystallisation, compatible elements such as Ni, Cr, V, Mg, and Co are most enriched in primitive magnetite (associated with Fe-Ti-V deposits) and decrease up-section towards the most evolved magnetite (associated with Fe-Ti-P deposits, Barnes et al. 2004;

Klemme et al 2006; Tegner et al. 2006; Namur et al. 2010; Dare et al. 2014). Within both titanomagnetite- and hemo-ilmenite-dominated oxide deposits hosted in anorthosite, magnetite has also been used as an indicator of primary magmatic conditions. Within the Grader Lake (hemo-ilmenite dominated) Ti deposit, Québec, Charlier et al. (2008) noted that Ni and Cr contents within Ti-poor magnetite decreased up-section while the concentrations of major elements (i.e. Ti, Al, Mg, Mn) remained relatively constant. Within the titanomagnetite dominated oxide deposits of the Suwalki anorthosite, evolution-related trends show that Cr and Mg within titanomagnetite decreased up-section whereas Mn increased up section (Charlier et al. 2009). Charlier et al. (2009) also noted that low V within titanomagnetite was related to high $fO_2$ of the parental magma. He et al. (2016) linked the more primitive composition (elevated Ni, Cr, and V) of magnetite within massive ores (massive oxide and nelsonite) relative to the (nelsonitic) gabbronorite unit at Damiao, China as an indicator that the massive ore bodies formed from a different magma than the host gabbronorite and anorthosite, indicating immiscible separation between Fe-Ti-P-rich and $SiO_2$-rich magmas. In most of these previous studies, the effect of sub-solidus exsolution on the magnetite chemistry were assumed to be negligible as the effects of diffusive modification via re-equilibration with silicates were not evaluated.

According to Pang et al. (2007), titanomagnetite within the Panzhihua layered intrusion (host of the world's largest V mine) was modified by 3 different processes; 1) re-equilibration between magnetite and ilmenite (inter-oxide re-equilibration), 2) re-equilibration between Fe-oxides and Fe-Mg silicates (oxide-silicate re-equilibration), and 3) re-equilibration between magnetite and internal exsolution phases (intra-oxide re-equilibration). Inter-oxide re-equilibration involves the exchange of 1) $Fe^{2+}$ and $Ti^{4+}$ from magnetite with

$Fe^{3+}$ in ilmenite and 2) $Fe^{2+}$ in ilmenite for $Mg^{2+}$ in magnetite (equation 1). Oxide-silicate re-equilibration involves the exchange of $Mg^{2+}$ in magnetite for $Fe^{2+}$ in ferromagnesian silicates (equation 2), predominantly olivine and pyroxenes (Frost et al. 1988; Frost 1991; Frost and Lindsley 1992). Intra-oxide re-equilibration involves the removal of Ti from Ti-magnetite through oxy-exsolution of the ulvöspinel component of magnetite to form ilmenite (equation 3; Frost 1991).

$$Equation\ 1: Fe^{2+}(in\ ilmenite) + Mg^{2+}(in\ magnetite)$$
$$\leftrightarrow Fe^{2+}(in\ magnetite) + Mg^{2+}(in\ ilmenite)$$
$$Equation\ 2: Mg^{2+}(in\ Fe-oxides) + Fe^{2+}(in\ Fe-Mg\ silicates)$$
$$\leftrightarrow Mg^{2+}(in\ Fe-Mg\ silicates) + Fe^{2+}(in\ Fe-oxides)$$
$$Equation\ 3: 6Fe_2TiO_4(ulvospinel) + O_2 \leftrightarrow 6FeTiO_3(ilmenite) + 2Fe_3O_4(magnetite)$$

All of the described reactions proceed to the right upon cooling (Frost et al. 1988; Frost 1991; Frost and Lindsley 1992), producing Fe-oxide minerals with compositions closer to their pure end member (with the exception of Mg enrichments in ilmenite) and Fe-Mg silicates with compositions closer to their Mg-rich end member. Within massive Fe-oxide samples, the reactions displayed by equations 1 and 3 will have the strongest effects on Fe-oxide composition whereas in samples containing semi-massive/disseminated Fe-oxides, reactions described by all 3 equations will have affected Fe-oxide composition. Therefore, 1) exsolution and/or re-equilibration with ilmenite and/or Al-spinel and 2) re-equilibration with co-existing silicate phases must be considered before the interpretation of magnetite composition can be used to interpret magmatic crystallization conditions.

Based on the above, it is likely that magnetite has undergone significant modification from its primary composition. Therefore, it is necessary to evaluate the effect of post-magmatic processes on the chemistry of Fe-oxides. Given that the Lac St. Jean anorthosite suite has crystallization ages ranging from 1170-1140 Ma (Higgins and van Breeman 1992), and is hosted within the allochthonous MP belt (amphibolite to granulite facies metamorphism) of the Grenville Province (Rivers et al. 2012), the anorthositic intrusions likely cooled over a long period of time. Therefore, it is important to evaluate if the composition of magnetite still can be a reliable indicator of primary magmatic conditions.

### 9.1.2 Modification of magnetite composition by exsolution and re-equilibration of Fe-oxides

#### 9.1.2.1 Sub-solidus exsolution processes

Magnetite has the formula $AB_2O_4$ where A represents the tetrahedral site containing $Fe^{3+}$ and B represents the octahedral site carrying $Fe^{2+}$ and $Fe^{3+}$. Divalent cations, such as $Mg^{2+}$, can substitute for $Fe^{2+}$ in the B site and both trivalent and tetravalent cations, such as $Ti^{4+}$, $V^{3+}$, $Cr^{3+}$, and $Al^{3+}$, can substitute into the A and B sites. Titanomagnetite represents a solid solution series between magnetite ($Fe_3O_4$) and ulvöspinel ($Fe_2TiO_4$: Basta 1960; Webster and Bright; Taylor 1961; Lindsley 1962; Taylor 1964). When crystallizing from a mafic magma, a titanomagnetite crystal can contain approximately 17-25 wt.% $TiO_2$ (Buddington and Lindsley 1964) and <5 wt.% $Al_2O_3$ (within titanomagnetite at Panzhihua; see Pang et al. 2007). The preservation of ulvöspinel is relatively rare, because under relatively oxidizing conditions, ulvöspinel is unstable and changes to ilmenite via oxy-exsolution (Buddington and Lindsley 1964).

The quantity and textural patterns of exsolutions within magnetite are dependent on the 1) bulk composition of titanomagnetite crystals and 2) amount of sub-solidus re-equilibration which occurred (Price and Putnis 1979; Price 1980; Amcoff and Figuierdo 1990). The sub-solidus exsolution process that forms ilmenite or Al-spinel within magnetite is related to 1) solubility decrease of components within solid solution series, 2) supersaturation of solid solution components through rapid compositional changes/diffusion, and 3) oxy-exsolution of ulvöspinel forming ilmenite exsolutions (see review in Arguin et al. 2018). There are 4 different processes by which ilmenite exsolutions within magnetite can form; 1) oxy-exsolution of ulvöspinel at T<600°C, which forms ultra-fine, cloth-textured lamellae (Ramdohr 1953; Price 1980, 1981); 2) oxy-exsolution of ulvöspinel at T>600°C, which allows ample time for exsolutions to coarsen and form trellis-, sandwich-textured ilmenite exsolutions, or internal granules upon slow cooling (Buddington and Lindsley 1964; Haggerty 1991); 3) direct exsolution from cation-deficient solid solution, which is limited by the quantity of vacancies within a spinel crystal (Lattard 1995); and 4) extensive sub-solidus re-equilibration of co-existing Fe-oxides that can generate further ilmenite exsolution in magnetite and even magnetite exsolution in ilmenite (Tan et al. 2017). Ilmenite can also exsolve and migrate outside the host magnetite crystal at high temperatures through external granule exsolution (e.g., Von Gruenewaldt et al. 1985).

Exsolutions of Al-spinel (Fe, Mg, Al spinel) within magnetite can form at $T_{max}$ of approximately 900°C as Mg and Al decrease in solubility within magnetite upon cooling (Brown et al. 1957; Turnock and Eugster 1962; Buddington and Lindsley 1964). Ilmenite can also exsolve hematite-lamellae if 1) ilmenite is a liquidus mineral phase from a high-Ti magma and 2) high $fO_2$ of the magma allows the incorporation of approximately 7-9 wt.% hematite into the ilmenite structure (Balsey and Buddington 1958; Charlier et al. 2015).

*9.1.2.2   Origin of coarse granular ilmenite and spinel*

Reconstruction of primary (pre-exsolution) magnetite composition from the lower zone massive oxide unit of the Panzhihua intrusion by Pang et al. (2007), using the technique coined by Bowles (1977) and Frost and Chacko (1989) involving careful integration of exsolved phases (ilmenite + Al-spinel) within titanomagnetite, gives compositions similar to that of the bulk rock composition of the corresponding massive oxide ore (approximately 16 wt.% $TiO_2$, 4.9% wt.% $Al_2O_3$, 5.4 wt.% MgO; Zhou et al. 2005). From this they inferred that ilmenite was secondary in origin, generated via exsolutions rather than primary in origin (i.e., crystallized directly from the magma). If ilmenite was indeed a primary magmatic phase, one should also see elevated Ti/Fe ratios within whole rock analysis of massive Fe-oxide ore relative to a massive Fe-oxide sample containing mostly titanomagnetite (Song et al. 2013; Howarth et al. 2013).

The origin of coarse-grained ilmenite and Al-spinel within oxide mineralization hosted by the Lac St. Jean anorthosite suite was also evaluated using bulk rock Ti and Fe contents of all samples examined in this study (Fig. 8.3A). These samples are compared with bulk-rock compositions of 1) massive oxides from the Lac Tio Ti deposit (Charlier et al. 2008), which is dominated by hemo-ilmenite, and 2) Buttercup Fe-Ti-V deposit (this study), which contains the most elevated modal percentage of titanomagnetite and lowest amount of coarse-exsolutions (5-10% Ilm, 1-5% spinel) within titanomagnetite relative to the other deposits examined in this study. Samples of massive oxides from all deposits hosted within the Lac St. Jean anorthosite suite in this study display similar whole-rock Ti-Fe compositions to that of Buttercup even though they contain variable amounts of coarse-grained ilmenite (ranging from 20-50 modal % ilmenite). Therefore, coarse-grained ilmenite probably formed through external granule exsolution of titanomagnetite, rather than by direct crystallization. It is most likely that coarse-grained spinel also formed through external granule exsolution from magnetite through a similar process to ilmenite whereby Mg and Al diffuse outside magnetite to form spinel. Semi-massive and disseminated samples examined in this study also plot along the Buttercup magnetite trend indicating that their bulk composition is related to the modal proportion of original titanomagnetite (approximately 21.9% TiO2 value at 69.27% $Fe_2O_{3(tot)}$).

If ilmenite was a liquidus phase, the bulk-rock Ti contents should be elevated and plot on a steeper Ti/Fe trend, closer towards that of Lac Tio. This is because under oxidizing conditions, where crystallization of magnetite is favored over ilmenite, ilmenite can crystallize if the magma is adequately enriched in $TiO_2$ (Toplis and Carrol 1995). The abundance of hemo-ilmenite, containing exsolutions of hematite, within anorthosite-hosted deposits (i.e. Lac Tio Fe-Ti deposit, Québec: Charlier et al. 2008) indicates that ilmenite can crystallize under oxidizing conditions if the magma was Ti-rich. Therefore, it is likely that the parental magmas associated with mineralization within the Lac St. Jean anorthosite suite contained insufficient $TiO_2$ to crystallize (hemo)-ilmenite as a liquidus phase.

In contrast to the oxide mineralization hosted within the Lac St. Jean anorthosite suite, it is likely that (hemo)-ilmenite was a liquidus mineral phase during fractional crystallization within the Lac Vanel-hosted Lac à l'Orignal Fe-Ti-P occurrence, despite low whole rock Ti/Fe ratios of massive oxides (Fig. 8.3A). This is based on 3 lines of evidence; 1) the dominance of hemo-ilmenite (> 10 wt.% $Fe_2O_3$) within nelsonitic norites, 2) the presence of significant Ti and HFSE depletions within magnetite relative to other deposits examined in this study (Fig. 6.8F) due to competition with co-crystallising ilmenite, and 3) Mo partitions preferentially into ilmenite at Lac à l'Orignal whereas at the other deposits examined in this study it preferentially partitions into magnetite (Figs. 6.8F and 6.9E). Ilmenite from the massive oxides at Lac à l'Orignal contains elevated $Fe^{3+}$ contents (8 wt.% $Fe_2O_3$) relative to ilmenite from the other deposits. However, since no exsolution phases are visible, it cannot be classified as hemo-ilmenite. This may be due to extensive re-equilibration between magnetite and ilmenite which permitted magnetite to strip sufficient $Fe^{3+}$ from ilmenite to supress formation of hematite exsolutions within the host ilmenite. Low whole rock Ti/Fe ratios of massive oxide ore from Lac à l'Orignal should contradict the interpretation of ilmenite as the sole liquidus mineral. However, magnetite was also likely a liquidus mineral phase as indicated by the associated massive oxide samples from Lac à l'Orignal. It could be possible that significant amounts of low-Ti magnetite included in the whole rock analysis could cause low bulk-rock Ti contents. Therefore, hemo-ilmenite was likely the first liquidus mineral and was subsequently followed by crystallization of Ti-poor magnetite, similar to what is observed in the Grader Lake Fe-Ti-P deposit (see Charlier et al. 2008).

### 9.1.2.3 *Summary of Fe-oxide exsolution textures and trends*

In order to interpret the effect of exsolution on magnetite chemistry, a summary of exsolution textures for the different oxide deposits/occurrence is given as follows. Coarse-grained ilmenite and Al-spinel are interpreted as external granule exsolutions because 1) the growth of smaller ilmenite and Al-spinel crystals tends to take the shape of grain boundaries between magnetite crystals whether they are between 2 crystals or at triple junctions and 2) ilmenite contains diffusive crystal faces when in contact with magnetite (see Fig. 3.4C and D; Fig. 3.10B and C). External granules of ilmenite and Al-spinel are the most common exsolution texture within the deposits examined in this study. However, modal percentages of external granules of ilmenite and Al-spinel differ among the deposits examined in this study and also at the deposit scale (see Figs 3.4A, 3.4C & D; Figs. 4.4A-C). For example, external granules of ilmenite and Al-spinel are most abundant at Lac Margane/Houlière-Nord (approximately 40-50% ilmenite and 10-20% Al-spinel) but generally rare (<10% ilmenite + Al-spinel) at Buttercup. However, massive oxide ore from Buttercup generally shows an elevated quantity of external granules of ilmenite and spinel within samples taken at the contact (approximately 10-20% ilmenite and 5-10% Al-spinel) with the host anorthosite relative to samples taken away from the contact. At the Buttercup deposit, the increasing percentage of external granules of ilmenite and Al-spinel related to proximity to the contact can be correlated with decreasing magnetite Ti contents (i.e. 17.99-17.10 wt.% $TiO_2$ away from contact, 18.35-7.97 wt.% $TiO_2$ at contact). The Lac Margane/Houlière-Nord massive oxide, which displays significant external granule exsolution of ilmenite and Al-spinel, contains magnetite that is significantly depleted in Ti (0.43 wt.% $TiO_2$) relative to Buttercup massive oxide.

The most common internal exsolution textures within magnetite from the study area (summarized in Table 9.1) are (in order of decreasing abundance); 1) trellis-textured Al-spinel, 2) fine-grained granular Al-spinel, 3) trellis-textured ilmenite, 4) sandwich-textured ilmenite, and 5) ultra-fine cloth-textured ilmenite. Trellis-textured and fine-grained granular Al-spinel exsolutions (and absence of internal ilmenite exsolutions) are generally associated with low-Ti magnetite (<2.5 wt.% $TiO_2$) whereas trellis-textured and sandwich textured ilmenite exsolutions are associated with high Ti-magnetite (>2.5 wt.% $TiO_2$); ultra-fine cloth-textured ilmenite exsolutions are associated with the highest Ti contents (>14 wt.% $TiO_2$) in magnetite (Table 9.1). Decreasing Ti contents within magnetite and its relation to exsolution type is likely correlated with an increasing degree of oxy-exsolution and sub-solidus re-equilibration with Fe-oxide phases. Exsolutions of hematite within ilmenite (hemo-ilmenite) are only visible within the nelsonitic norite unit from the Lac à l'Orignal occurrence, which is associated with magnetite containing negligible Ti contents (<0.05 wt.% $TiO_2$). In contrast, the massive oxide unit at Lac à l'Orignal contains exsolution-free ilmenite and Ti-poor (<1 wt. % $TiO_2$) magnetite with trellis lamellae of Al-spinel.

The habit and modality of exsolutions within magnetite are also dependent on the modal percentage of oxides to silicates (i.e. massive vs. disseminated; see Chapters 3 and 4). Generally, a modal decrease in Fe-oxides from massive to disseminated ores is correlated with 1) increasing modal proportions of granular ilmenite to magnetite, 2) decreasing modality of all exsolution phases within magnetite, 3) disappearance of coarse Al-spinel crystals and 4) decreasing Ti contents of magnetite. This is best seen within the Lac Margane/Houlière-Nord and Lac à Paul mineralization. This also indicates an increase in degree of sub-solidus re-equilibration with coexisting mineral phases (Fe-oxides and silicates).

**Table 9.1:** Summary of exsolution textures of Fe-oxide minerals within the 6 Fe-Ti-V-P deposits/occurrences examined in this study. Cumulus implies primary crystallization rather than formation through external granule exsolution. Mt=magnetite, spl=spinel, U-F=ultra-fine (<5 µm), F-gr=fine-grained, msv.= massive.

| | | | | Exsolution Textures | | | | | | | | |
| | | | | External (crystals>0.5cm) | | Internal | | | | | | |
| | | | | | | Within magnetite (arranged from left to right in order of increasing grain size) | | | | | Within ilmenite | |
| Host anorthosite | Deposit Type | Deposit Name | Oxide assemblage | Granular ilm | Granular spl | U-F cloth-textured ilm | F-gr granular spinel | Trellis lamellae spinel | Trellis lamelae ilm | Sandwich ilm | Hematite lamellae | |
|---|---|---|---|---|---|---|---|---|---|---|---|---|
| Lac St. Jean (1170 - 1140 Ma) | Fe-Ti-V | Buttercup | mt>>ilm | * | * | X | X | X | X(near contact) | X | | |
| | | Lac-Margane/ Houlière-Nord | mt + ilm | X | X | | X | X | | | | |
| | Fe-Ti-P | St. Charles de Bourget | mt>ilm | X | X | X(in nelsonite) | X | X | X | X | | |
| | | Lac Perron | mt+ilm | X | X | | X | X | X | X | | |
| | | Lac à Paul | ilm>mt | X | X | | X | X | X | | | |
| Lac Vanel (1080 ± 2 Ma) | Fe-Ti-P | Lac à l'Orignal (msv. oxide) | ilm + mt | cumulus | X | | | X | | | | |
| | | Lac à l'Orignal (nelsonitic norite) | hemo-ilm>mt | cumulus | X | | | X | | | X | |

X= Common          * = Less common

Visible evidence of sub-solidus re-equilibration is noted by 1) the presence of exsolution-rich cores and exsolution free rims within magnetite in contact with ilmenite grains (Figs. 3.3C and D, 3.10A-C, 4.4B) and 2) diffusive crystal faces of ilmenite in contact with a magnetite grain (Figs. 3.4D, 3.5B, 3.10A-D, 4.4B, 4.10C, 4.13A). The diffusive crystal faces could also be an overgrowth of later exsolutions, which migrate onto pre-existing ilmenite. Within magnetite crystals containing ultra-fine-cloth-textured exsolutions (i.e. magnetite from massive oxides at Buttercup and nelsonite from St. Charles de Bourget), no visible evidence of sub-solidus re-equilibration is present. This indicates that magnetite underwent a very low degree of external granule exsolution or sub-solidus re-equilibration with other Fe-oxide phases. Visible evidence of sub-solidus re-equilibration within ilmenite crystals is only visible at Lac à l'Orignal by the thinning and gradual disappearance of hematite lamellae within ilmenite towards the contact with magnetite grains (Fig. 4.15B).

### 9.1.2.4 *Effect of exsolution processes on in-situ magnetite chemistry*

The effect of the exsolution process on bulk-rock compositions is believed to be negligible as the formation of exsolutions should just re-distribute elements among the different oxide minerals. Therefore, bulk rock trace element compositions of massive Fe-oxide samples can be used as a potential baseline to 1) evaluate how the chemistry of magnetite, determined by the LA-ICP-MS, can be modified by variable amounts of exsolution on the grain scale and 2) constrain the composition of primary Ti-magnetite before exsolution occurred. This is based on the interpretation that the original massive oxide was composed solely of aluminous-rich titanomagnetite and that all ilmenite and spinel has crystallized through exsolution, as argued in section 8.2.3. Whole rock compositions of massive Fe-oxide samples from all deposits in this study, except for Lac à Paul (massive oxides not available in this study), are compared to the in-situ analyses of titanomagnetite obtained by LA-ICP-MS in

Fig. 9.1 and clearly shows that the exsolution process has altered the primary compositions of magnetite.

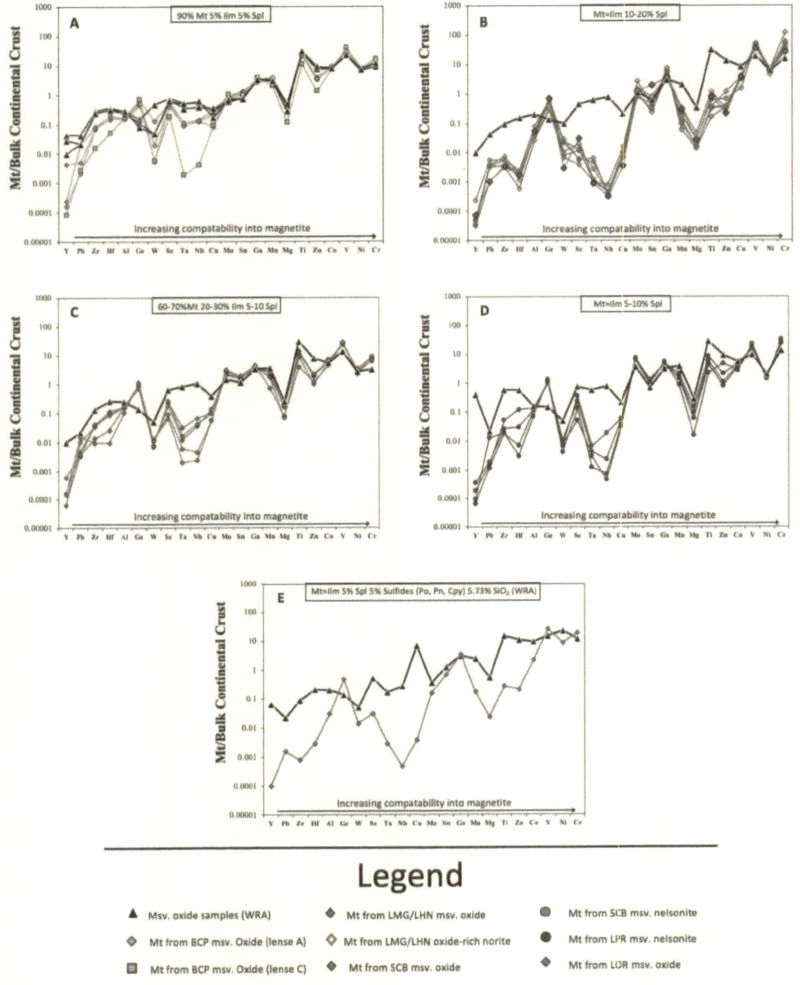



## Legend



**Figure 9.1:** Multi-element variation diagrams, normalized to bulk continental crust, for whole rock analysis (WRA) of massive oxides (black lines) compared to magnetite composition obtained by LA-ICP-MS. A) Buttercup (BCP) Fe-Ti-V, B) Lac Margane/Houlière-Nord Fe-Ti-V, C) St. Charles de Bourget Fe-Ti-V-P, D) Lac Perron Fe-Ti-P, E) Lac à l'Orignal Fe-Ti-P. Text within the top center rectangle of each plots represents the modal proportions of mineral phases within the representative massive oxide sample. Mt=magnetite, ilm=ilmenite, spl=spinel, Po=pyrrhotite, Pn=pentlandite, cpy=chalcopyrite, msv=massive, BCP=Buttercup, LMG=Lac Margane, LHN=Lac Houlière-Nord, SCB=St. Charles de Bourget, LPR=Lac Perron, LOR=Lac à l'Orignal.

The differences in the distribution of elements in whole-rock massive oxides relative to distribution of elements within magnetite, determined by LA-ICP-MS, can be related to the degree of which ilmenite and spinel have exsolved from magnetite. This is based on the general enrichment of Cr, V, and Ge (preferentially partition into magnetite) and the depletion of Ti, Mg, Mn, Zn, Co, and HFSE (preferentially partition into ilmenite and Al-spinel) in magnetite relative to the patterns displayed by the whole-rock compositions of massive Fe-oxide ores (Fig. 9.1). Exsolution of spinel is of lower significance due to low modal percentages within samples and lower Al and Mg contents within the whole-rock values of massive Fe-oxide samples but, will likely result in magnetite being depleted in Mg, Al, Zn, and Co and enriched in Cr and V.

Magnetite from the Buttercup deposit, which contains the lowest modal proportion of ilmenite exsolutions relative to the other deposits appears to display the composition which has been modified the least (Fig. 9.1A). Among the other deposits, in order of increasing modification of the trace element composition of magnetite, due to exsolution of ilmenite and spinel, can be arranged as follows: 1) St. Charles de Bourget (Fig. 9.1C); 3) Lac Perron (Fig. 9.1D); and 4) Lac Margane/Houlière-Nord (Fig. 9.1B). Lac à l'Orignal (Fig. 9.1E) was not included in this ranking because, as described earlier, ilmenite is believed to be a primary igneous phase rather than an external granule exsolution, but the competition of elements between ilmenite and magnetite should be relatively similar. All samples from Lac à Paul were too $SiO_2$-rich (>6 wt.%) to make an accurate interpretation of the effects of exsolution on magnetite chemistry based on comparison with whole rock massive oxide composition, and is not presented here.

Knowing that the composition of magnetite has been modified significantly in most cases due to exsolutions (Fig. 9.1), it is necessary to examine if the whole rock composition of massive oxides (originally titanomagnetite) can be a reliable indicator of primary magmatic conditions, as it is possible that indicators of magmatic evolution within magnetite (V, Ni, and Cr) have been enriched to varying degrees due to exsolution of ilmenite and Al-spinel, as clearly shown at Lac Margane/Houlière-Nord. Therefore, whole-rock composition of massive oxides may be a more reliable indicator of magmatic evolution in cases where it can be proven that it comprises titanomagnetite alone.

The comparison of massive oxide whole rock data in Fig. 9.2 shows that massive oxide from the Fe-Ti-V deposits/occurrences (Buttercup and Lac-Houlière-Nord) is generally enriched in Ni, Cr, and V relative to Fe-Ti-P deposits/occurrences (St. Charles de Bourget, Lac Perron and Lac à l'Orignal; Fig. 9.2), correlating with increasingly fractionated magma forming Fe-Ti-P deposits. The exception is massive oxide from Lac Perron which contains anomalously high Cr for a Fe-Ti-P deposit for reasons currently unknown. Massive oxide from Buttercup and Lac Margane/Houlière-Nord display similar Mg, Ti, Co, V, and Cr contents which indicates that magnetite composition between these two locations was relatively similar before exsolution occurred. However, massive oxide from Buttercup appears to be more enriched in HFSE relative to Lac Margane/Houlière-Nord. Elevated concentrations of chalcophile elements (Ni, Co, Cu) within massive oxide from Lac à l'Orignal (Fig. 9.2) could indicate a more primitive composition but, is more likely caused by the presence of magmatic sulfides (approximately 5% pyrrhotite + pentlandite + chalcopyrite) within the massive oxide sample. Therefore, based on the patterns shown in Figure 9.2, the oxide mineralization can be ordered in increasing degree of magmatic evolution as: 1) Lac

Margane/Houlière-Nord Fe-Ti-V, 2) Buttercup Fe-Ti-V, 3) Lac Perron Fe-Ti-P, and 4) St. Charles de Bourget Fe-Ti-V-P. Ranking the massive oxide sample from Lac à l'Orignal Fe-Ti-P occurrence in terms of magmatic evolution based on Figure 9.2 is difficult, even though V and Cr contents are similar to Fe-Ti-V mineralization of Buttercup and Lac Margane, due to 1) its location within a younger anorthosite suite, 2) depleted whole rock Ti (12.05 % $TiO_2$; Fig. 8.3A and Fig. 9.2), Nb, and Ta contents relative to the other deposits/ occurrences, and 3) the presence of sulfides.

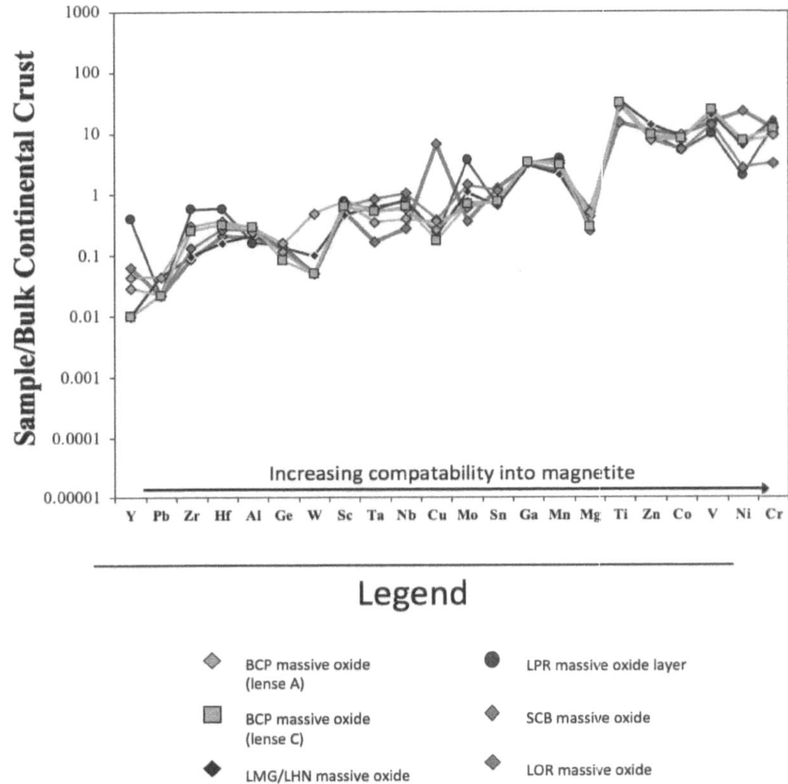



**Figure 9.2:** Multi-element variation diagrams, normalized to bulk continental crust, for whole rock massive oxides (SiO$_2$<7%) compared to magnetite composition obtained in situ by LA-ICP-MS from all deposits examined in this study. BCP=Buttercup, LMG=Lac Margane, LHN=Lac Houlière-Nord, LPR=Lac Perron, SCB=St. Charles de Bourget, LOR=Lac à l'Orignal.

Both LA-ICP-MS data and whole-rock data are plotted on binary diagrams of key elements (Ni, Cr, V) which are commonly used to track the degree of fractionation (Fig. 9.3). The whole-rock values for massive oxides are also plotted on Figure 9.3 to examine if the laser data remains a robust tracer of fractionation even when extensive exsolution takes place. The laser data shows that magnetite from Fe-Ti-V mineralization was elevated in Ni, Cr, and V and therefore, as expected, formed from a more primitive melt than magnetite from Fe-Ti-P mineralization (Fig. 9.3A and B). Whole rock analysis of massive oxides from the Buttercup Fe-Ti-V deposit and the Lac Margane/Houlière-Nord Fe-Ti-V occurrence show similar whole rock contents of Ni, Cr, V, and Ti whereas LA-ICP-MS analysis of magnetite shows elevated Cr and V from Lac Margane/Houlière-Nord relative to Buttercup (Fig. 9.3). Also, magnetite (analysed by LA-ICP-MS) from massive oxides associated with Fe-Ti-P mineralization at St. Charles de Bourget and Lac à l'Orignal have Ni, Cr and V contents which plot within the Fe-Ti-V field while magnetite from apatite-bearing samples (nelsonite at St. Charles de Bourget and nelsonitic norite at Lac à l'Orignal; Fig. 9.3) plot within the Fe-Ti-P field. However, the whole rock contents of Ni, Cr, and V from these massive oxide samples at St. Charles de Bourget and Lac à l'Orignal are not as enriched as the laser data and plot in the Fe-Ti-P field. This indicates that the exsolution of ilmenite and Al-spinel in from magnetite should lead to an increase in Cr and V with simultaneous decrease in Ti and little change in Ni concentrations within magnetite (Fig. 9.3). This shows that the degree of fractionation using Cr and V alone can be misinterpreted if only LA-ICP-MS data of magnetite is used for samples where magnetite has undergone extensive exsolution of ilmenite and Al-spinel, such is the case at Lac Margane/Houlière-Nord. Therefore, exsolution processes which increase Cr, and V contents within magnetite could possibly create a magnetite with a more primitive composition even if initial magmatic compositions are similar.

Concentrations of Ni in magnetite (laser data) and bulk rock massive Fe-oxide analyses appear to be relatively similar. Therefore, within massive Fe-oxide samples, Ni in magnetite appears to be a robust indicator of magmatic evolution in spite of the compositional changes caused by significant exsolution of ilmenite and Al-spinel. However, the Ni content of massive Fe-oxide samples obtained by whole-rock is of course influenced by the presence of magmatic sulfides. This is the case for Lac à l'Orignal where the massive Fe-oxide ore contains approximately 5% sulfide minerals, including pentlandite, which explains the bulk-rock enrichment in Ni relative to other deposits examined in this study (Fig. 9.3C). Therefore, the Ni-Cr diagram is the most robust indicator for magmatic evolution using laser data of magnetite.

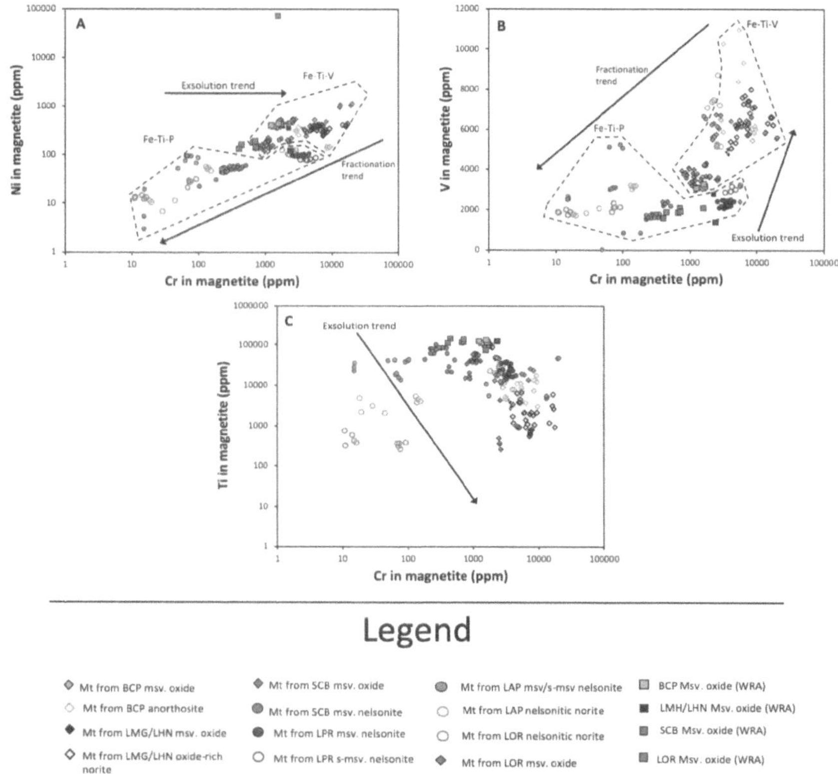



**Figure 9.3:** Binary plots of trace element compositions of magnetite (LA-ICP-MS) and whole-rock analysis (WRA) of massive oxides (square symbols). A) Ni vs. Cr, B) V vs. Cr, C) Ti vs Cr. Arrows depicting exsolution trend approximated based on change in composition between whole rock massive oxide and corresponding magnetite. Mt=magnetite, msv=massive, s-msv=semi-massive, BCP=Buttercup, LMG=Lac Margane, LHN=Lac Houlière-Nord, SCB=St. Charles de Bourget, LPR=Lac Perron, LAP=Lac à Paul, LOR=Lac à l'Orignal.

### 9.1.3   Modification of magnetite composition by re-equilibration with silicates

Previous literature has shown that during prolonged cooling of an intrusion a sub-solidus exchange of elements occurs between Fe-oxides and Fe-Mg silicates (i.e. olivine and othopyroxene; Morse 1980; Frost et al. 1988; Frost 1991; Frost and Lindsley 1992; Pang et al. 2007; Charlier et al. 2015). This reaction results in the loss of Mg from magnetite into the Fe-Mg silicate minerals creating false enrichments in Mg (e.g., Charlier et al. 2015). This was first noted in the Kiglapait intrusion where Ti-magnetite lost Mg in the presence of olivine and augite (Morse et al. 1980). Sub-solidus re-equilibration between Fe-oxides and co-existing silicates occurs by a process called diffusive modification, described by Tanner et al. (2014) as igneous crystals obtaining equilibrium with the adjacent media (melts, minerals, or volatile-rich fluids). The rate at which diffusive modification occurs depends on: 1) Ionic radius and charge of the element (ionic potential) - ions with a larger ionic radius and lower charge (i.e. $Mg^{2+}$, $Li^{+}$) will move more freely than ions with a small ionic radius and high charge (i.e. HFSE); 2) Diffusion coefficient for an element into a mineral; 3) Duration of cooling; 4) Temperature; 5) Compositional gradient; 6) Adjacent mineral composition; 7) Oxygen fugacity - $fO_2$; 8) Partition coefficients of an element into a mineral; 9) $aSiO_2$; 10) Diffusion pathways (Chackraborty 2008; Costa et al 2008; Faak et al. 2013). Specifically, for Fe-oxides and Fe-Mg silicates, the pyroxene-QUILF geothermometer (created by Andersen et al. 1993) indicates diffusive modification can only occur at temperatures above 950°C (Pang et al. 2007).

Tanner et al. (2014) studied the effects of diffusive modification within silicates from the unmetamorphosed Bushveld Complex, and determined that diffusive modification had affected the composition of silicate minerals during cooling of this large layered intrusion. Diffusive modification had occurred through 4 different processes; 1) intra-granular diffusion within zoned crystals, 2) inter-granular diffusion, 3) large-scale diffusive processes, and 4) trapped liquid shift effect. Intra-granular diffusion is indicated by the absence of oscillatory zoning within minerals (Tanner et al. 2014). Inter-granular diffusion is indicated by an offset in chemical composition between different minerals, for example, Fe-Mg silicates displaying elevated Mg# than expected due to absorption of Mg from another Mg-bearing mineral (Tanner et al. 2014). Large-scale diffusive processes are difficult to find direct evidence of, but hydrous fluids can possibly enhance the rate of diffusion within silicate minerals (Liu and Yund 1992), Therefore, one should look for hydrous mineral phases as potential evidence of this process. Trapped liquid shift involves the reaction of cumulus minerals with an evolved intercumulus liquid, enriched in incompatible elements. The resulting mineral compositions will reflect the ratio of cumulus to intercumulus phases rather than primary magmatic composition (Barnes 1986; Cawthorn 2007). However, the trapped liquid shift effect will only cause significant changes to the composition of a mineral if the mineral in question is in very low abundance (<5%) and a large portion of trapped liquid (>10%) is present (see trapped liquid shift calculations with apatite by Cawthorn 2007). These processes effect the primary magmatic compositions of accessory cumulus minerals to the point where their resulting compositions reflect composition after completion of sub-solidus re-equilibration, rather than composition after crystallization. It is imperative that these diffusive modification processes are considered before interpreting compositions of Fe-oxide minerals as reflective of their primary magmatic signature.

*9.1.3.1    Visible evidence for diffusive modification in the Lac St. Jean anorthosite-hosted*
*deposits*

Unlike magmatic oxide mineralization from the Bushveld Complex, magmatic oxides from the Lac St. Jean anorthosite have been subject to amphibolite-granulite facies metamorphism related to the Grenville orogeny. Elevated temperatures associated with metamorphism could also create the conditions necessary for significant diffusive modification to occur but it would be difficult to distinguish between chemical changes related to sub-solidus re-equilibration during cooling and sub-solidus re-equilibration during metamorphism without possessing unmetamorphsed samples for comparison. Therefore, the compositional changes in magnetite in the presence of silicates are interpreted to result from 1) initial high-T (>950 °C) diffusive modification between Fe-oxides and Fe-Mg silicates during cooling of the intrusion followed by 2) interactions between Fe-oxides and silicates during amphibolite-granulite facies metamorphism (<950 °C).

Interactions between Fe-oxides and silicate minerals during amphibolite-granulite facies metamorphism (i.e. <950 °C) can be noted by the presence of amphibole $\pm$ biotite $\pm$ garnet coronas surrounding Fe-oxides. These textures were first noted within oxide-bearing olivine metagabbros by Whitney and McLelland (1983) who determined that coronas around Fe-oxides formed through a 2-stage reaction (Reactions 1 and 2) at granulite facies conditions. However, these coronas were described as poorly developed or absent within olivine-free rocks because the quantity of Mg diffusing out of Fe-oxides was deemed inefficient and needed to originate from the breakdown of Fe-Mg silicates (Reaction 1; Whitney and McLelland 1983). Both reactions can result in the loss of Ca from plagioclase crystals creating the appearance of a more evolved composition (lower An content).

$$Reaction\ 1: Plagioclase + Fe - Ti\ oxides + Olivine + water$$

$$\leftrightarrow amphibole + Al - spinel + orthopyroxene +\ biotite$$

$$+ more\ sodic\ Plagioclase$$

$$Reaction\ 2: amphibole + orthopyroxene + Al - spinel + plagioclase$$

$$\leftrightarrow Garnet + clinopyroxene + more\ sodic\ plagioclase + water$$

Petrographic examination of all samples from each deposit/occurrence shows visible evidence of sub-solidus re-equilibration between Fe-oxides and silicate minerals; it is most commonly present within samples that contain disseminated and semi-massive Fe-oxides. However, it can be present locally within massive samples. Within disseminated Fe-oxide samples examined in this study (mostly Fe-oxide-bearing anorthosites from Buttercup, St. Charles de Bourget, Lac Perron, and Lac à l'Orignal), Fe-oxide minerals in contact with plagioclase crystals are locally surrounded either by mm-scale coronas of garnet, or coronas of biotite ± amphibole (Figs. 3.5D and F, 4.10G, and 4.11D).

Within semi-massive ore samples (e.g., Fe-oxide norites from Lac Margane/Houlière-Nord and semi-massive nelsonite from Lac Perron), these coronas are generally thicker (>0.5 cm), predominantly composed of an inner amphibole layer and an outer biotite layer. Coronas of garnet occur locally and often contain inclusions of amphibole and/or biotite. Within the semi-massive dunitic nelsonites from Lac à Paul New Zone, coarse-grained olivine crystals in contact with Fe-oxides are surrounded by a μm-scale corona of orthopyroxene (Fig 4.13D). The enriched source of Mg from olivine and lack of hydrous fluids in this case possibly limited reaction 1 and in the process prevented reaction 2 from occurring.

Visible evidence for sub-solidus re-equilibration between Fe-oxides and Fe-Mg silicates within massive samples is only noted locally within massive oxide samples from

Buttercup. These re-equilibration textures are highly complex as they consist of symplectites of olivine and nepheline surrounded by a corona of high-Ti amphibole (%). In most cases the symplectic olivine and nepheline is rarely preserved and only the high-Ti amphibole is present. Diffusive modification between Fe-oxides and olivine in this case was significant as the olivine symplectites can display Fo contents of $Fo_{71}$ indicating significant transfer of Mg from Fe-oxides into olivine.

Most of the host lithologies of the Fe-Ti-V-P deposits/occurrences examined in this project are anorthosites (e.g., Buttercup, Lac Perron) or leuco/gabbronorite (e.g, Lac Margane/Houlière-Nord, Lac à Paul New Zone, Lac à l'Orignal). Olivine-bearing lithologies are less common and orthopyroxene is only present in significant quantities within host rocks from Lac Margane/Houlière-Nord, Lac à Paul, and Lac à l'Orignal. Therefore, the source of Mg for Fe-Mg silicate-poor deposits is relatively difficult to constrain. The source of $H_2O$ in these reactions is also difficult to constrain but is possibly related to the volatile content of the parental magma. Despite the lack of an obvious Mg and hydrous fluid source in anorthosite-hosted oxide mineralization, the coronas generally appear to be relatively well-developed despite lacking production of ortho- and clinopyroxene by reactions 1 and 2 respectively. Also, the increased thickness of the coronas and increased modification of co-existing primary silicate minerals can be correlated with the increasing modal proportions of Fe-oxides (i.e. disseminated to massive). Although Whitney and McLelland (1983) determined that the breakdown of Fe-oxides is inadequate in producing the required Mg for these reactions, this conclusion is based on examination of disseminated oxide samples. Therefore, it is possible that increasing the quantity of primary magnetite, which contain elevated MgO (>2% in magnetite; i.e. massive oxides at Buttercup), could increase the production of amphibole and/or biotite in Reaction 1. The increasing presence of Fe-oxides could also be responsible for elevated Fo values within olivine symplectites from Buttercup due to the increased surface area for diffusion. The formation of garnet coronas surrounding Fe-oxides likely resulted from a lack of $H_2O$ which limited the formation of amphibole and biotite. This would permit more rapid consumption of these minerals in order to proceed to (Reaction 2) more rapidly.

### 9.1.3.2 *Effect of re-equilibration with silicates on magnetite chemistry*

In order to constrain how the re-equilibration of Fe-oxides with Fe-Mg silicates affects the composition of magnetite, the trace element contents of several cations within magnetite are plotted against Mg, which readily exchanges between Fe-oxides and Fe-Mg silicates during diffusive modification (Fig. 9.4). This is done to investigate the differences in composition between magnetite within massive samples to that of disseminated samples. However, these elements can also be affected by the process of (oxy-)exsolution of ilmenite (i.e. Ti, Mg) and/or spinel (i.e. Mg, Al, Zn) as well as sub-solidus re-equilibration with silicates. Therefore, it is important to try to constrain which trace elements within magnetite will be depleted during sub-solidus re-equilibration. For all deposits/occurrences, magnetite from disseminated samples is generally depleted in Ti (Fig. 9.4A), Mn (Fig. 9.4B), and Al (Fig. 9.4C) and slightly enriched in V (Fig. 9.4D) and Cr (Fig. 9.3) relative to magnetite from massive samples. On a log scale, Ni appears to be relatively unaffected by sub-solidus re-equilibration or exsolutions as its concentrations within magnetite does not to appear to change significantly during sub-solidus re-equilibration (Fig. 9.4E). The only noticeable change in Zn contents of magnetite are at Buttercup, where magnetite from anorthosites contain enriched Zn relative to magnetite from massive oxides (Fig. 9.4F).

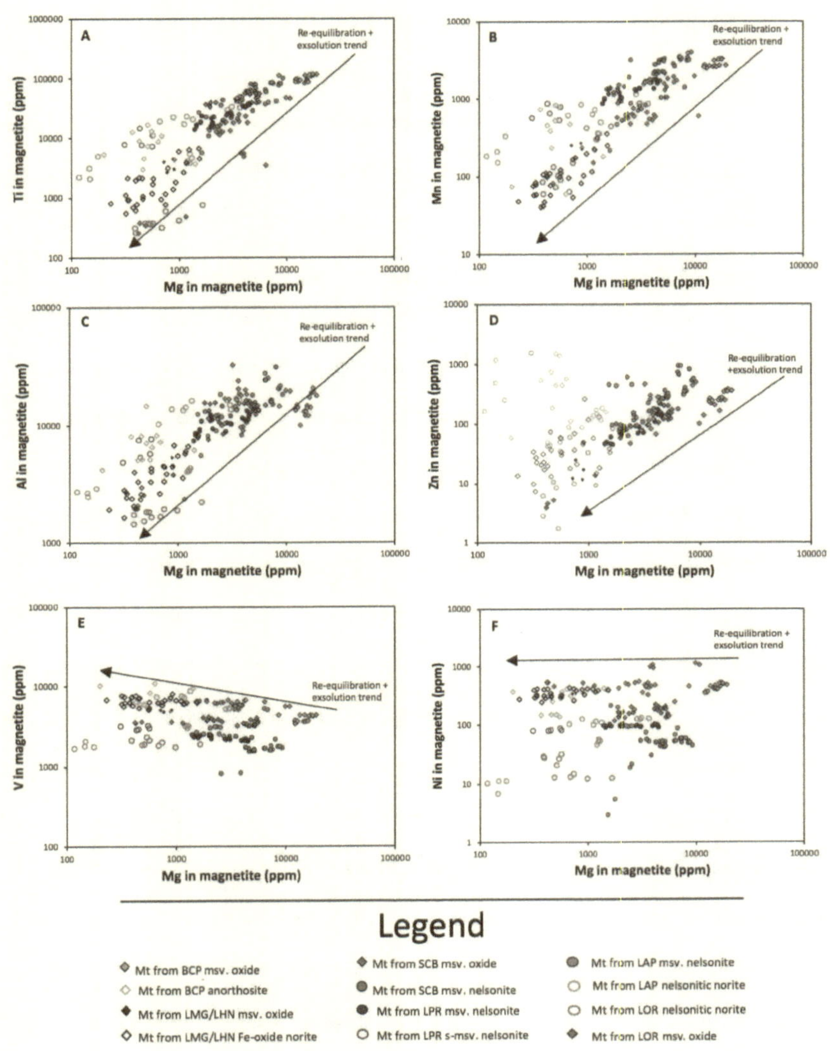



**Figure 9.4**: Binary plots showing the effects of diffusive modification (re-equilibration with silicates) and (oxy-) exsolution of ilmenite and Al-spinel on magnetite composition. A) Ti vs. Mg, B) Mn vs. Mg, C) Al vs. Mg, D) Zn vs. Mg, E) V vs. Mg, F) Ni vs. Mg. Mt=magnetite, msv=massive, s-msv=semi-massive, BCP=Buttercup, LMG=Lac Margane, LHN=Lac Houlière-Nord, SCB=St. Charles de Bourget, LPR=Lac Perron, LAP=Lac à Paul, LOR=Lac à l'Orignal.

184

In order to constrain whether elements within magnetite become enriched or depleted as a result of concurrent exsolution and sub-solidus re-equilibration, we consider the following; 1) oxy-exsolution of ilmenite from magnetite depletes the resulting magnetite in Mg, Mn, Ti, and HFSE, 2) exsolution of Al-spinel from magnetite depletes the resulting magnetite in Mg, Zn, and Al, 3) sub-solidus re-equilibration with Fe-Mg silicates depletes the resulting magnetite in Mg, Mn, and Al (if re-equilibrating with amphibole or biotite), 4) sub-solidus re-equilibration between magnetite and plagioclase should also deplete the resulting magnetite in Al, and 5) all of the previously described processes will create relative enrichments in the Cr and V contents in magnetite due to the reduction of magnetite volume after extensive exsolution. Surprisingly, Ni is least affected. The process of exsolving ilmenite from magnetite occurs more readily and extensively within disseminated samples. However, Al-spinel is relatively absent within disseminated oxides compared to massive oxides. We suggest that this is evidence for diffusion of Al out of magnetite during sub-solidus re-equilibration as it likely 1) enters plagioclase, amphibole, or biotite or 2) forms Al-spinel crystals, which then become consumed during the corona-forming reactions.

In summary, magnetite from disseminated samples is much less representative of its primary composition during crystallization than magnetite from massive samples. This is due to the combination of 1) increased efficiency of oxy-exsolution, 2) sub-solidus re-equilibration with silicates by diffusive modification, and 3) corona-forming reactions. The combined effects of sub-solidus re-equilibration, corona-forming reactions and (oxy-) exsolution on magnetite composition has modified its primary composition to a much greater extent than magnetite from massive samples. The higher degree of ilmenite exsolution in disseminated oxides compared to massive oxides has also been reported at Panzhihua Fe-Ti-V deposit and is possibly due to increased availability of $O_2$ within samples containing disseminated Fe-oxides (see Pang et al. 2007).

Magnetite from massive oxide samples away from the contact from the Buttercup Fe-Ti-V deposit would thus be most representative of primary magmatic composition in the Lac St. Jean anorthosite as magnetite displays little compositional change from its whole-rock values. This means that the magnetite has undergone little compositional changes due to exsolution of ilmenite and Al-spinel and since silicate minerals are not present, the modification of Mg and other divalent cations will be limited. On contrast, magnetite from the massive oxides at Lac Margane/Houlière-Nord (Fe-Ti-V) is significantly depleted in Mg relative to magnetite from massive samples at Buttercup. In this case, the increased departure of Mg and other divalent cations from magnetite is directly correlated with the increased degree of exsolution of ilmenite and Al-spinel at Lac Margane/Houlière-Nord relative to Buttercup. However, the depletion of Mg within magnetite from disseminated samples relative to that in massive samples at Lac Margane/Houlière-Nord is significantly less relative to other deposits. This could indicate that the exsolution of ilmenite and Al-spinel from magnetite is more effective in altering the primary composition of magnetite than sub-solidus re-equilibration with silicates.

## 9.2 Evaluating apatite chemistry as a petrogenetic indicator of anorthosite-hosted mineralization

Igneous apatite is an abundantly common accessory mineral which can crystallize within many different rock types (Watson, 1980). However, in this study of Fe-Ti-V-P rich rocks, apatite is commonly highly abundant (>5 modal%). Crystallization of apatite is controlled predominantly by the composition of the magma; as P is highly incompatible in silicate minerals it thus behaves incompatibly during fractional crystallization (Cawthorn and Walsh 1988; Toplis et al. 1994; Tollari et al. 2008). Therefore, only within the most evolved cumulates within layered mafic intrusions is apatite present as a cumulus mineral phase (see Namur et al. 2010). The composition of igneous apatite is mainly controlled by the composition and fO$_2$ of the magma and partition coefficients ($D_{melt}^{Ap}$) for REE increase with increasing polymerization of the melt (Prowatke and Klemme 2006). Within mafic-intermediate magmas, apatite will likely control not only the volatile element concentrations (i.e. F, Cl) but also REE, U, and Th budgets in whole rock compositions as it is the only mineral which contains significant amounts of these elements (Prowatke and Klemme 2006; Tollari et al 2008). Therefore, within the samples examined in this study, the REE content of apatite, in particular, should be an excellent indicator of 1) magmatic evolution, 2) depth/degree of partial melting from the mantle source, 3) phases involved in fractional crystallization, 4) fO$_2$ of the magma, and 5) information on the parental melts which formed Fe-Ti-P deposits. However, like with Fe-oxides, the possible effects of secondary processes such as on the composition of apatite must be examined first.

### 9.2.1 Possible post-cumulus changes in apatite and its composition

Unlike with Fe-oxides, the composition of apatite cannot be modified through sub-solidus exsolution. There is also no clear physical evidence of sub-solidus re-equilibration with co-existing mineral phases. For example, corona reaction textures akin to those seen surrounding Fe-oxides are not present around apatite crystals. Apatite examined in this study is present within both nelsonites, at Lac Perron and St. Charles de Bourget, and more disseminated mineralized lithologies, such as nelsonitic norites at Lac à Paul area (the ore zones and New Zone in the footwall) and Lac à l'Orignal, and nelsonitic ultramafics (specific to Lac à Paul area). Therefore, one must consider post-crystallization processes during comparison of apatite compositions between differing lithologies. Apatite is generally considered to be resistant to post-cumulus compositional changes relative to other major rock-forming minerals (Cherniak 2010). However, since the composition of Fe-oxides have been modified significantly post-crystallization, it makes sense to examine also the potential processes that may also modify the composition of apatite. Four processes are examined; 1) textural coarsening, 2) sub-solidus re-equilibration through diffusive modification, 3) trapped liquid shift effect, and 4) metamorphism.

### 9.2.1.1  Textural coarsening of apatite

Apatite crystals are usually present as clusters of subhedral to anhedral crystals, which can locally contain micro- and macroscopic inclusions of Fe-oxides (Fig. 4.5D, 4.9B, 4.10E and F). However, the grain size of apatite can be highly variable among each deposit/occurrence in the Lac St. Jean area; 1) apatite from massive nelsonite at Lac Perron is very coarse-grained (cm-scale, Figs. 4.9B, 4.10A, 4.10E, and F), 2) apatite from St. Charles de Bourget (massive nelsonite, Figs. 4.5A and 4.5C-E) and Lac à l'Orignal (nelsonitic norite Figs; 4.15C and E) is relatively medium-grained (approximately 500 µm), and 3) apatite from mineralized samples occurring in the footwall (New Zone) of Lac à Paul is very fine-grained (<100 µm, Figs. 4.13D-F). Apatite from Lac Perron is not only significantly coarse-grained, but also displays more irregular-shaped (anhedral) crystals with coarser inclusions of Fe-oxides (>0.5cm) relative to apatite from the other deposits examined in this study. Therefore, we propose below that apatite from Lac Perron has undergone significant textural coarsening.

The coarsening of a mineral during slow cooling of an intrusion or metamorphism is called textural coarsening, or Otswald Ripening, and occurs through a process by which smaller crystals dissolve with the simultaneous growth of larger crystals (Higgins 2011). One crystal will impinge onto another crystal of the same mineral and coarsen until the crystals change into a more equant form (Higgins 2011). Otswald ripening should not affect whole rock composition as the grain boundaries of mineral phases are just being re-distributed. This process has been described within troctolitic rocks from the southwest corner of Lac St. Jean anorthosite suite (100-200 km south of Lac Perron) whereby very large olivine oikocrysts (up to 30 cm) contain inclusions of fine-grained plagioclase that were protected from coarsening (Fig. 9.5A; Higgins 1998). This is similar to the textures shown by apatite within nelsonite from Lac Perron (Fig. 9.5B). Textural coarsening has also been noted within other well-studied mafic intrusions, such as Kiglapait, Labrador (Higgins 2002) and Skaergaard, Greenland (Higgins 2011). Experimental studies have determined that olivine grows significantly faster than plagioclase during Ostwald ripening because olivine follows the screw-dislocation-limited $t^{1/3}$ growth trend and plagioclase follows the surface nucleation-limited log(t) growth trend (Cabane et al. 2005). Experiments performed by Muller et al. (1999) indicate that apatite also follows the $t^{1/3}$ growth law. Therefore, apatite should coarsen significantly during Otswald ripening.

**Figure 9.5:** Images comparing the textures obtained through textural coarsening. A) Figure from Higgins (1998) showing very coarse-grained (>20 cm) olivine crystals with inclusions of plagioclase in the anorthosite of Lac-St. Jean. B) Outcrop photograph showing very-coarse-grained apatite crystals containing inclusions of Fe-oxides (nelsonite ore from Lac Perron).

According to Higgins (2011), there are many factors that determine the rate of growth during Otswald ripening. Those relevant for this study include; 1) the mineral in question (apatite), 2) temperature of re-equilibration or metamorphism, 3) duration of increased temperature period, 4) grain size of apatite, and 5) modal percentage of apatite. Although we do not have any quantitative constraints on temperature or duration of crystallization and/or metamorphism, it is assumed that if mineral phases are very coarse-grained relative to other phases, then elevated temperatures persisted over a longer period of time. Therefore, all interpretations on Ostwald ripening are qualitative. Nelsonite from Lac Perron was then possibly subject to elevated temperatures for a longer time period due to slower cooling rates and/or elevated degrees of metamorphism since apatite grains there are the coarsest (i.e. similar conditions to nodular troctolite examined by Higgins 1998). Although nelsonite from St. Charles de Bourget is the richest in apatite, Otswald ripening was not as significant in the growth of apatite as that at Lac Perron. This could likely be due to either 1) limited space of apatite impingement due to the presence of olivine xenoliths within the nelsonite at St Charles de Bourget, whereas nelsonite from Lac Perron is pure nelsonite (only Fe-oxides and apatite) or 2) shorter time period of elevated temperatures. The former is the likely reason for the lack of coarsening of apatite within the Lac à l'Orignal deposit due to the abundance of other primary mineral phases relative to apatite (hemo-ilmenite, magnetite, plagioclase, orthopyroxene). Apatite from the footwall (New Zone) of the Lac à Paul Fe-Ti-P deposit is likely very fine-grained possibly due to lower overall proportions of apatite within the 1) nelsonitic ultramafics and 2) nelsonitic norites. Low modal proportions of apatite would mean that grains could not impinge onto another grain efficiently enough to cause significant coarsening.

The effects of Otswald Ripening on the chemistry of apatites analyzed in this study are difficult to constrain. However, larger apatite grains (i.e. significantly coarser than the 50μm beam diameter) are preferred for the method of LA-ICP-MS analysis used in this study because with larger grains, it is easier to avoid visible inclusions while analyzing linear patterns within a crystal. Therefore, if a coarser apatite grain is analyzed, there is a higher degree of confidence that the results will be more representative of a relatively pure apatite grain without inclusions.

### 9.2.1.2  *Sub-solidus re-equilibration*

The crystal structure of apatite (2 cation sites, M1 and M2) allow a wide variety of substitutions to take place for Ca. Divalent cations can substitute directly for Ca while trivalent and tetravalent cations substitute for Ca though coupled substitution (Hughes and Rakovan 2015). Substitutions can also occur in the anion site as F, OH, and Cl can readily substitute for one another (Hughes and Rakovan 2015). Therefore, it is possible that any sub-solidus substitution reactions could take place during periods of elevated temperatures (Cherniak 2010): 1) Volatile elements diffuse the most rapidly; 2) Sr, LREE, and Mn all diffuse out of apatite at similar rates; 3) Diffusion would happen most readily at T>1000 °C; 4) Coupled substitution would require temperatures >800 °C at 1 atm; 5) The effects of pressure on diffusion remain unclear. Apatite cores have been known to retain magmatic Mn contents at T=700 °C on a million-year timescale (Cherniak 2005). However, due to the duration of magmatism within the Lac St. Jean anorthosite suite (1170-1140 million years old) it is possible that Mn contents within apatite could have been modified over this time. Since Al-spinel exsolves from magnetite at approximately $T_{Max}$=900 °C (Tan et al. 2017), these conditions could be favorable to produce compositional changes in apatite due to diffusive modification with coexisting mineral phases, such as Fe-oxides and Fe-Mg silicates.

Possible evidence of sub-solidus re-equilibration of apatite with silicates is shown by a general observation that apatite is relatively depleted in Na, Mn, and Mg within disseminated mineralization at Lac à Paul and Lac à l'Orignal relative to apatite from massive nelsonite at St. Charles de Bourget and Lac Perron (Figs. 7.4, 7.5). It is therefore possible that apatite could have lost Na to plagioclase and Mn and Mg to Fe-Mg silicates. If this trend is related to diffusive modification then we should expect a positive correlation of these elements with Cl, since Cl is displaced from apatite during re-equilibration with volatile phases, especially with increasing metamorphic conditions (Lieftink et al. 1994). In the case of this study, such a correlation is not present. One should also consider that these apatites are from different deposits so these geochemical trends could be related to differing magmatic conditions or magma compositions. Therefore, it is best to compare apatite composition from massive and disseminated mineralization from the same deposit/occurrence. Unfortunately for this study, in most of the P deposits both mineralization styles do no co-exist. However, in the Lac à Paul footwall (New Zone), it is possible to compare apatite composition from ultramafic nelsonites (considered semi-massive: > 40% Fe-oxides + apatite with olivine ±orthopyroxene) and nelsonitic norites (disseminated: <15% Fe-oxides + apatite with plagioclase and orthopyroxene). There are no significant compositional changes in apatite (Cl, Mg, Mn, Sr, or REE) with respect to the abundance of silicates (Figs. 7.4, 7.5A-C). Therefore, sub-solidus re-equilibration is not considered to be a significant process affecting magmatic apatite compositions.

### 9.2.1.3 Trapped liquid shift effect

The concentration of an element within apatite after re-equilibration with trapped liquid is dependent on 1) its primary magmatic concentration within apatite, 2) modal proportions of all cumulus minerals, 3) partition coefficients for an element into cumulus phases, 4) proportion of trapped liquid - which is noted to increase up stratigraphy, and 5) composition of the trapped liquid (Cawthorn 2013). Calculations by Cawthorn (2013) showed the important effect of trapped liquid on modifying the LREE content of accessory (<5 modal %) apatite in the Bushveld Complex, South Africa. Within the Merensky Reef lower zone, he calculated that 5% of intercumulus apatite, with an original Ce concentration of 330 ppm, re-equilibrated with 10% trapped liquid to increase Ce contents of apatite to 521 ppm. Similarly, for the upper zone, 2% cumulus apatite re-equilibrated with 25% trapped liquid to increase the Ce concentration in apatite from 330 ppm to 1762 ppm.

According to Cawthorn (2013), compositional modification of apatite due to re-equilibration with up to 30% trapped liquid only has significant effects when the normative apatite is below (4%). Although, for the samples examined in this study, apatite is consistently present in modal percentages $\geq$10 %. Therefore, the effects of re-equilibration with trapped liquid on apatite composition should be minimal. This statement has been proven true by Charlier et al. (2005) who stated that cumulus apatite buffers REE distribution among cumulus phases and will control REE contents within bulk cumulates. Re-equilibration between apatite and trapped liquid should also create large negative Eu anomalies as Eu will be re-distributed between both apatite and plagioclase (Cawthorn 2013). The Eu anomalies shown within apatite examined in this study are relatively shallow (Eu/Eu*$\geq$0.6). Therefore, apatite and trapped liquid did not re-equilibrate to a significant degree, an observation also shared by She et al. (2016) upon examination of apatite within apatite-bearing gabbronorites in Damiao, China.

### 9.2.1.4 *Metamorphism*

Studies on apatite within the Sudbury Igneous Complex (SIC) show that apatite volatile and LREE contents are susceptible to change during metamorphism (Warner et al.

194

1998): Apatite within the amphibolite portion of the SIC are depleted in LREE and Cl relative to apatite from the greenschist portion. The Cl content of apatite is likely replaced by OH or F (Lieftink et al. 1994). However, the effect of metamorphism on apatite in this study is difficult to determine because metamorphic grades of individual deposits were not quantitatively identified for comparison. However, they are expected to range from granulite to amphibolite facies, due to their location within the allochthonous MP-LP belt (Rivers et al. 2012) and the abundance of garnet, amphibole, and biotite as secondary phases. Therefore, it is assumed that apatite within each Fe-Ti-P deposit/occurrence examined experienced similar grades of metamorphism and only its duration could potentially affect LREE and Cl loss in apatite.

9.2.2   Interpreting magmatic conditions using apatite REE abundance

Based on the above section, it is assumed that apatite compositions are more accurate indicators of primary magmatic conditions than Fe-oxide minerals. Previous geochemical analysis of apatite within Fe-Ti-P deposits in layered intrusions (e.g., Grader Lake, anorthosite-hosted: Charlier et al. 2005; Sept Iles layered intrusions: Tollari et al. 2008) shows that REE contents (except Eu) increase up sequence during fractionation. Thus, the most evolved apatite should contain the most elevated REE contents. However, the increase in REE within apatite should not affect the slope of chondrite-normalized plots unless crystallization of another REE-bearing phase occurs. The most elevated total REE contents within apatite are found within Lac à l'Orignal (1819 - 3505 ppm) followed by St. Charles de Bourget (2452 - 2960 ppm), then Lac Perron (1952 -2834 ppm), and the most primitive apatite composition is present at Lac à Paul (1092-2419 ppm). The concentration of Cl within apatite also appears to be a good indicator of magmatic evolution as it is positively correlated with total REE contents of apatite (Fig. 7.5). This data shows that the Lac à l'Orignal deposit crystallized from the most evolved magma and the Lac à Paul deposit crystallized from the most primitive magma assuming that apatite is the main host of REE contents within each lithology.

To determine whether apatite is the main host of REE contents within apatite the relative abundances of LREE to HREE in apatite must be examined. LREE/HREE ratios within apatite are remain relatively similar within each deposit except for Lac à Paul. This potentially indicates early saturation of an HREE-bearing phase during fractionation. Mass balance calculations performed on ore samples at Lac à Paul by Chartier-Montreuil et al. (2017) indicates apatite is the main host of REE content but that < 20% of the whole rock HREE budget is likely hosted by amphibole and pyroxene. Since HREE are moderately incompatible in clinopyroxene $D_{Gd-Lu}^{CPX/Liq} = 0.62 - 0.71$ and orthopyroxene $D_{Gd-Lu}^{OPX/Liq} = 0.052 - 0.30$ whereas LREE are strongly incompatible in clinopyroxene $D_{La-Eu}^{CPX/Liq} = 0.05 - 0.46$ and orthopyroxene $D_{La-Eu}^{OPX/Liq} = 0.00042 - 0.034$ (Hart and Dunn 1993; Hauri et al. 1994; Lee et al. 2007), fractionation of either pyroxenes or amphibole, before or during apatite crystallization, should be considered.

Although HREE within apatite at Lac à Paul appear to become progressively depleted relative to LREE up stratigraphy from the ultramafic nelsonites to the nelsonitic norites, at Lac à l'Orignal the LREE/HREE ratios of apatite remain relatively constant, where orthopyroxene is also present in significant quantities. This indicates that pyroxenes are likely not responsible for HREE depletions in apatite. Instead, it could be possible that crystallization of an accessory mineral before or during fractionation, such as zircon where $D_{HREE}^{Zircon/melt} > D_{LREE}^{Zircon/melt}$ (Rubatto and Hermann 2007), could have been responsible for HREE depletions in apatite. Although a detailed search for zircon in the mineralization was out of the scope for this project, zircon was not observed by petrography or during EMPA in all samples.

However, ferrodiorites from Lac à Paul are elevated in Zr compared to ferrodiorites from St. Charles de Bourget (Appendix 7). More work should be done to confirm the role of zircon or other accessory minerals involved with REE fractionation processes within these deposits.

Using the obtained results, it is possible to examine whether these deposits crystallized from a similar magma at different stages of evolution or from a different mantle source altogether. According to Winter (2010), elevated LREE/HREE and MREE/HREE ratios, which indicate HREE depletions, are evidence of either 1) a lower degree partial melt or 2) a deeper melting source which has garnet in the residual melt. However, HREE content of apatite could also be affected by co-crystallization of HREE-bearing minerals, e.g. zircon, as discussed above for Lac à Paul. Based on relative abundances of HREE to LREE within apatite based on chondrite-normalized (Fig. 7.2) and LREE/HREE (Fig. 7.3) the following model is proposed, which is tested in Section 9.2.5:

1) Apatite from Lac Perron and St. Charles de Bourget display relatively similar and constant LREE/HREE abundances and probably formed from a similar magma.

2) Apatite from peridotitic nelsonites from Lac à Paul ore zone display similar LREE/HREE to Lac Perron and St. Charles de Bourget but displays an HREE-depleted signature up stratigraphy within the nelsonitic norites, indicating that the Lac à Paul deposit formed from a similar magma to Lac Perron and St. Charles de Bourget but HREE become gradually depleted due to zircon crystallization.

3) Finally, apatite from Lac à l'Orignal displays the highest LREE/HREE ratios relative to the other deposits examined in this study, indicating that apatite crystallized from a parental magma of different composition.

9.2.3   Interpreting fO₂ conditions using apatite trace element composition

Constraining the fO₂ of a magma using apatite could be performed using two methods, 1) Eu/Eu* of apatite (see Charlier et al. 2008) and 2) Mn content of apatite (applied to felsic rocks -see Miles et al. 2014). The $D^{Ap}_{Melt}$ value for Eu is sensitive to fO₂ since $Eu^{3+}$ is more compatible into apatite (Roelandts and Duchesne 1979) whereas $Eu^{2+}$ is more compatible into plagioclase (Drake and Weill 1975; Aigner-Torres et al. 2007). Therefore, 1) the Eu content of apatite should increase and 2) the negative Eu anomaly on chondrite-normalized plots should become shallower (Eu/Eu* increase) with increasing fO₂. The concentration of Mn within apatite can possibly be sensitive to fO₂ since $Mn^{2+}$ is more compatible into apatite than $Mn^{3+}$ because of its similar ionic radius to $Ca^{2+}$(Miles et al. 2014). However, Stokes et al. (2019) noted that polymerization of the melt has a much stronger effect on Mn concentration within apatite than fO₂. Therefore, Mn in apatite may increase with decreasing fO₂ and therefore, display a negative correlation with Eu/Eu*.

The Eu/Eu* values (Fig. 9.6) are highest, i.e. less of a negative Eu anomaly, within apatite from the Lac à l'Orignal deposit (Eu/Eu*= 0.72-1.16) compared to those from the other deposits/occurrences examined in this study (Eu/Eu*= 0.47-0.93). This, along with the presence of hemo-ilmenite as the dominant Fe-oxide phase, indicates that the Lac à l'Orignal Fe-Ti-P mineralization crystallized from a relatively oxidized and evolved magma relative to the other mineralization associated with the Lac St. Jean anorthosite. The combination of shallow Eu/Eu* values and presence of hemo-ilmenite was also noted at the Tellnes Ti deposit and also indicated crystallization from a relatively oxidized magma (Charlier et al. 2008). Enrichments in Mn contents of apatite within Lac Perron and St. Charles de Bourget relative to other deposits could also indicate that nelsonite ores crystallized in a more reduced environment. However, the lacking negative correlation Mn and Eu/Eu* indicates that in this case, Mn in apatite may not be a strong indicator of $fO_2$ conditions of the magma as indicated previously by (Stokes et al. 2019). This is possibly because within semi-massive-disseminated samples, Mn could substitute into Fe-Mg silicates during fractionation. Therefore, Eu/Eu* within apatite is considered to be a more reliable indicator of $fO_2$ conditions.

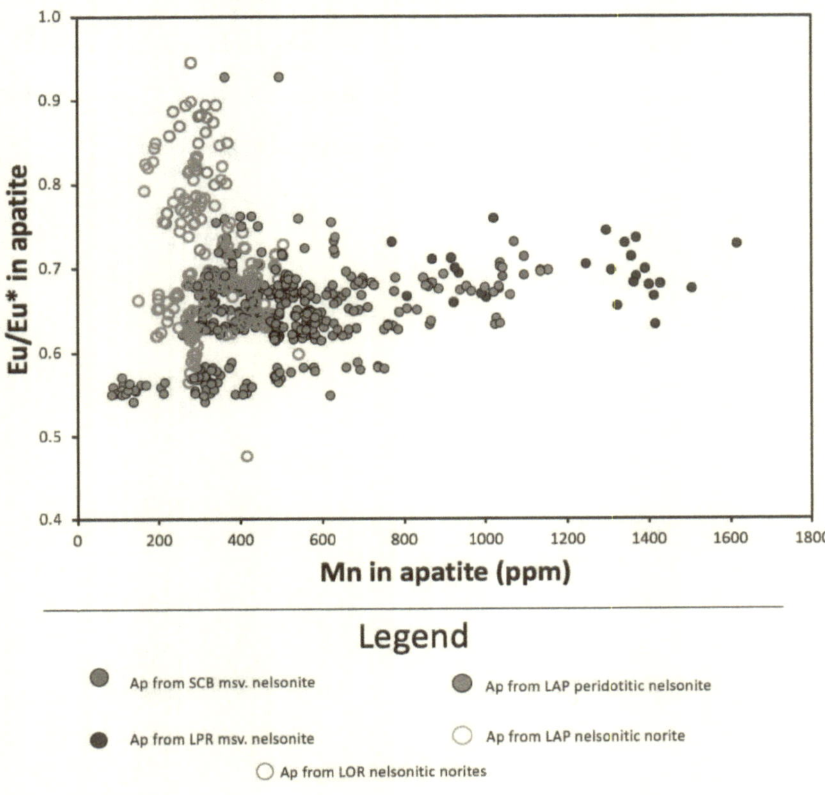


**Figure 9.6:** Binary plot of Eu/Eu* vs. Mn contents of apatite used to qualitively evaluate the $fO_2$ content of the magma. SCB=St. Charles de Bourget, msv= massive, LPR=Lac Perron, LAP=Lac à Paul, LOR=Lac à l'Orignal. LAP Apatite data supplemented by Chartier-Montreuil et al. (2017).

9.2.4   Parental magma composition

Previous work has shown that magmatic Fe-Ti-P deposits in anorthosite massifs are derived from fractionation of an evolved ferrodioritic magma (see Duchesne et al. 1989; Mitchell et al. 1996; Vander Auwera et al. 1998). Genetic links between ferrodoritic magma and Fe-Ti-P deposits have been noted through the comparison of calculated melt compositions using apatite with the composition of associated ferrodiorite dykes (see Dymek et al. 2001 - several AMCG-related intrusions in Québec and New York; Charlier et al 2005 - Bjerkreim-Sokndal intrusion in Norway; Charlier et al. 2008 - Grader Lake intrusion in Québec; and Tollari et al. 2008 - Sept Iles Layered intrusion in Québec). In all cases, compositions of parental magma were obtained using the inversion method, which involves dividing the concentration of an element within apatite ($C_{Ap}$) by its partition coefficient with apatite ($D_{Melt}^{Ap}$). However, the $D_{Melt}^{Ap}$ values used in each case were different because Charlier et al. (2005) calculated $D_{Melt}^{Ap}$ values which were representative of ferrodioritic magma while Tollari et al. (2008) obtained $D_{Melt}^{Ap}$ from literature (see Bedard 2001, Watson and Green 1981, and Prowatke and Klemme 2006) which were most representative of basaltic magmas. Therefore, both sets of partition coefficients were used to calculate parental melts using apatite. Then, calculated melt compositions, using each $D_{Melt}^{Ap}$ value, were compared with chondrite-normalized REE plots of ferrodiorites from St. Charles de Bourget and Lac à Paul to examine which best reproduces parental melt composition (Fig. 9.7).

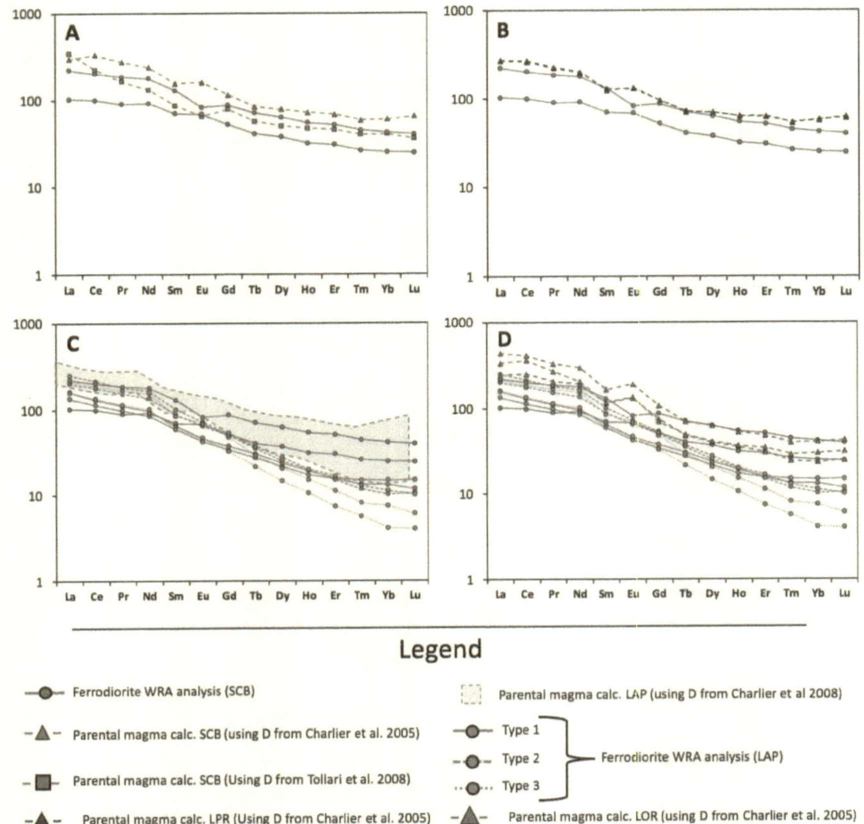



**Figure 9.7:** Chondrite-normalized REE plots displaying the calculated composition of parental melts using $D^{Ap}_{melt}$ values from Charlier et al. (2005) and Tollari et al. (2008) for Fe-Ti-P mineralization examined in this study, values from Sun and McDonough (1995). A) St. Charles de Bourget ferrodiorites and calculated parental melts, B) Lac Perron calculated parental melts plotted with St. Charles de Bourget ferrodiorites, C) Lac à Paul ferrodiorites, and Lac à Paul calculated parental melt field, plotted with St. Charles de Bourget ferrodiorites, D) Lac à l'Orignal parental calculated melts plotted with St. Charles de Bourget and Lac à Paul ferrodiorites. WRA=whole rock analysis, LAP=Lac à Paul, SCB=St. Charles de Bourget, LPR=Lac Perron, LOR=Lac à l'Orignal.

Figure 9.7A shows that chondrite-normalized REE abundances of calculated parental melts using Charlier et al. (2005) $D_{Melt}^{Ap}$ values display a sub-parallel pattern to corresponding ferrodiorites at the St. Charles de Bourget deposit. Chondrite-normalized REE abundances of calculated parental melt using Tollari et al. (2008) $D_{Melt}^{Ap}$ display an abnormal positive La anomaly relative to corresponding ferrodiorites at St. Charles de Bourget. (Fig. 9.7A). Therefore, parental magma calculations for ferrodiorites were best reproduced by using the $D_{Melt}^{Ap}$ values from Charlier et al. (2005) rather than those from Tollari et al. (2008). This is likely because the $D_{Melt}^{Ap}$ values used by Charlier et al (2005) were calculated specifically for ferrodioritic magmas. Only calculated melts using $D_{Melt}^{Ap}$ values used by Charlier et al (2005) are shown in Fig. 9.7B-D for clarity.

The chondrite-normalized slope for calculated parental melts at Lac Perron show similar slopes to chondrite-normalized slopes of ferrodiorites from St Charles de Bourget (Fig. 9.6B) confirming that the two deposits formed from magmas at 1) similar stages of evolution and 2) similar melt sources. However, the parental magma for St. Charles de Bourget may appear more evolved than the parental magma for Lac à Paul based on and 1) apatite total REE contents (Figs. 7.3, 7.5D, 7.7A-C) and 2) depleted Cr contents from whole-rock analyses of massive oxide samples (Figs. 9.1, 9.2, 9.3). Ferrodiorites from St. Charles de Bourget deposit display a much flatter slope on the chondrite-normalized plot than the field of calculated parental melts from Lac à Paul (Fig. 9.7C). However, in detail the St. Charles de Bourget ferrodiorite REE patterns match fairly well the upper limit of calculated field of LAP but not the lower limit of the field, which is HREE depleted. This indicates that the calculated parental melts from Lac à Paul are mostly HREE-depleted relative to the parental melts at St. Charles de Bourget. Ferrodiorite REE patterns from the Lac à Paul area are highly variable and do not match well with corresponding calculated melts (Fig. 9.7C). They are categorized according to the degree of HREE depletion relative to calculated parental melts at Lac à Paul: Type 1) sub-parallel to calculated melt field (low HREE depletion); Type 2) LREE-enriched with moderate HREE depletion; and Type 3) significantly HREE depleted (Fig 9.7C). These varying patterns may be caused by the presence or absence of co-crystallized HREE-bearing accessory mineral(s), e.g., zircon, as discussed in section 9.2.3, which was not considered in the calculation of parental melts. Therefore, parental melt compositions calculated from apatite might not fully represent parental melt if apatite is not the sole host of HREE.

At Lac à l'Orignal (hosted in the younger, Vanel anorthosite), the chondrite-normalized slope of calculated parental melts appears sub-parallel to Type 1 ferrodiorites from Lac à Paul, which is LREE enriched and moderately HREE-depleted (Fig. 9.7D). However, calculated parental melts from Lac à l'Orignal are significantly more enriched in REE overall relative to other calculated melts and ferrodiorites.

Comparison of all calculated parental melt compositions (Fig. 9.7) helps emphasise the points made in section 9.2.3, which is that these Fe-Ti-P deposits probably originated from three different magma sources based on the quantity of HREE depletions. 1) A magma with low degree HREE depletions relative to LREE formed the massive/semi-massive nelsonitic ores at Lac Perron and St. Charles de Bourget. 2) A similar magma but with possible early saturation of HREE-bearing accessory minerals (zircon), which causes the progressive depletion in HREE during magmatic evolution ($La_N/Yb_n$=6.5-17.9 in peridotitic nelsonites and $La_N/Yb_n$=2.5-25.6 in nelsonitic norites), formed the ores at Lac à Paul. Finally, 3) an overall LREE-enriched magma, with the most significant HREE depletions, formed nelsonitic norite at Lac à l'Orignal. It is expected that the parental melt would be significantly different at Lac à l'Orignal from the other 3 Fe-Ti-P deposits/occurrences as 1) it is hosted within the younger (1080 ± 2 Ma; van Breeman 2009) and much smaller Lac Vanel anorthosite and 2) contains hemo-ilmenite, which is absent in Lac St Jean anorthosite (1165-1135 Ma; Hébert et al. 2005). The origin of different magma sources involved in the formation of Fe-Ti-P deposits in the S/SW margin (magnetite-dominated mineralization of St. Charles de Bourget and Lac Perron) relative to those hosted within the northern margin (ilmenite-dominated mineralization of Lac à Paul) of the Lac St. Jean anorthosite is more complex. It could possibly relate to different magma pulses as magmatism within the Lac St. Jean anorthosite suite ranges from 1170 Ma to 1140 Ma (Hébert et al. 2005). However, geochronology data for anorthosites specific to those areas is limited.

## 9.3 Formation of Fe-Ti-V-P deposits within the Lac St. Jean anorthosite area

The deposits/occurrences examined in this study were originally placed into different categories based on their Fe-oxide assemblage; 1) magnetite-dominated type (Mt-dom; Buttercup, St. Charles de Bourget, and Lac Perron) 2) magnetite + ilmenite type (Mt+ ilm; Lac Margane/Houlière-Nord and Lac à Paul) and 3) hemo-ilmenite type (H-ilm; Lac à l'Orignal). However, new insights from the whole rock and laser data have shown that Mt-dom and Mt+ilm deposits/showings can be grouped together since 1) they are all hosted within the Geon-11 (1170-1140 Ma) Lac St. Jean anorthosite suite and 2) differences in ilmenite abundance was noted as due to degree of exsolution from titanomagnetite. Therefore, the most significant differences between the deposits/occurrences hosted within the Lac St. Jean anorthosite suite is the presence (Fe-Ti-P) or absence of apatite (Fe-Ti-V) and the degree of ilmenite exsolution from magnetite. Apatite is present in 1) significant amounts within mineralization at St. Charles de Bourget, Lac Perron, and Lac à Paul, 2) very rare at Lac Margane/Houlière-Nord, and 3) absent at Buttercup. In general, magnetite from Fe-Ti-P mineralization has an evolved composition relative to magnetite from Fe-Ti-V mineralization indicated by lower Ni and Cr contents, despite post cumulus changes due to exsolution and sub-solidus re-equilibration. Despite the differences in magnetite composition, apatite composition is relatively similar among each deposit/occurrence within the Lac St. Jean anorthosite suite. The only exception is for Lac à Paul, where chondrite-normalized apatite compositions initially display a shallow slope, then becomes progressively HREE-depleted during fractionation, possibly due to co-crystallization of zircon. However, more work needs to be done to evaluate possible fractionation of zircon or other possible HREE-bearing accessory minerals at Lac à Paul. This indicates that each deposit/occurrence examined in this study could have formed from similar magmas but differ due to fractional crystallization sequences.

In the Lac St. Jean deposits/occurrences, whole-rock Ti content of massive oxides indicates that magnetite enriched in Ti (>20 wt.%) and Al (>6 wt.%) was the sole-liquidus phase and that the varying degrees of ilmenite and spinel present were dependant on the degree of exsolution. Modal proportion and textural patterns of ilmenite exsolutions within magnetite is predominantly controlled by the primary Ti content of magnetite and $fO_2$ of the magma while those from spinel would be dependant mainly on primary Al and Mg contents of magnetite (see Tan et al. 2017; Arguin et al. 2018). Due to the elevated Ti content of magnetite relative to Mg and Al in both these deposit types and the $fO_2$ control, ilmenite often dominates over Al-spinel as an exsolution phase. Although, when discussing the increased degree of ilmenite exsolution within Mt + Ilm deposits, the availability of $O_2$-bearing hydrous fluids possibly interacting with Fe-oxides must be considered. Therefore, it is likely that mineralization containing increased quantities of ilmenite (i.e. Lac Margane/Houlière-Nord) were increasingly subjected to $O_2$-bearing hydrous fluids relative to mineralization with lower quantities of ilmenite (i.e. Buttercup).

Hemo-ilm deposits cannot be linked directly to Mt-dom or Mt+Ilm deposits by degree of exsolutions, host lithology, or apatite REE distribution and therefore, must be discussed separately. In general, magmatic Fe-oxide mineralization within the Grenville Province hemo-ilm mineralization is restricted to younger (<1160 Ma) andesine-type anorthosites and Ti-Mt type deposits are restricted to older (>1130 Ma) labradorite-type anorthosites (Hébert et al. 2005). This was proven to be the case for this study area, for example, the Lac Orignal hemo-ilm-apatite showing in the $1080 \pm 2$ Ma Lac Vanel anorthosite of the Pipmuacan AMCG complex (Higgins and van Breemen 2009). Based on the abundance of $Fe^{3+}$-rich ilmenite and hemo-ilmenite as liquidus phases and strong LREE-enrichment and HREE-depletions within apatite from the Lac à l'Orignal showing, this mineralization is proposed to have formed from a magma that is 1) $TiO_2$-rich (at least > 4 wt%; Vander Auwera et al. 1998), 2) relatively oxidized, and 3) generated from a lower degree partial melt which could possibly have been subjected to significant crustal contamination. This observation agrees with previous observations made by Owens et al. (1994) and Owens and Dymek (2001, 2005).

9.4   Unconstrained processes within genetic models

Due to the complex nature of magmatic oxide deposits, several questions regarding their formation and the concentration of ore minerals still remain debated, such as 1) were Fe-oxides $\pm$ apatite concentrated by liquid immiscibility or fractional crystallization and accumulation? 2) what is the reason behind cross-cutting relationships of mineralization with anorthosites? It would be unlikely that a pure Fe-Ti oxide $\pm$ apatite melt could have separated via liquid immiscibility as these magmas could not be produced at geologically reasonable temperatures (Lindsley and Epler 2017). For example, pure nelsonitic magmas were produced experimentally at 1450°C and 1 atm but, through addition of significant F and $H_2O$ as well as decreasing pressure, melting temperatures were only reduced to 1200°C (Wang et al. 2017). The role of immiscibility is highly debated as more recent experiments have produced Fe-P-O liquids (Mungall et al. 2018; Hou et al. 2018) under reasonable magmatic conditions. This does not exclude liquid immiscibility as an important process since fractionation in mafic intrusions can lead to immiscible separation of ferrodiorite and granitic magmas (Daly Gap) as evidenced by melt inclusions in apatite in layered intrusions (Charlier 2013). This can be followed by formation of nelsonites as cumulates from the ferrodiorite magma (Duchesne et al. 1999; Dymek and Owens 2001; Namur et al. 2012) and silicate minerals would be separated from the ore minerals through crystal sorting or floatation (Charlier et al. 2015). A cumulate origin for these deposits is supported by the whole-rock chemistry as it is predominantly controlled by the amount of Fe-oxides within the sample. The role of liquid immiscibility in the formation of magmatic oxide deposits within the Lac St. Jean anorthosite suite remains unknown but, should not be overlooked as an important process. For further work, one should look specifically at possible melt inclusions within apatites.

The St. Charles de Bourget deposit is the only deposit noted in this study where mineralization displays obvious cross-cutting relationships with host anorthosite (i.e. anorthosite xenoliths within the ore and sheared contact between massive oxides and anorthosite). Lindsley and Epler (2017) proposed that in order for oxide ore bodies to cross-cut the host anorthosite, Fe-oxides would have to be mobile in some way. Therefore, it is important to consider the potential mobility of Fe-oxides in a sub-solidus state. Fe-oxide cumulates can potentially be squeezed out during emplacement of anorthosite through filter-press compaction (see Charlier et al. 2015), residual silicate liquid can provide lubrication for this process (Lindsley and Epler 2017). Lindsley and Epler (2017) also proposed that external granule exsolution of ilmenite can weaken Fe-oxides during recrystallization and increase their ability to flow. Till and Moscowitz (2013) indicated that magnetite is mechanically weaker than plagioclase at a given T and therefore, likely to flow at elevated, sub-solidus temperatures. Mobility of Fe-oxides during metamorphism or deformation at elevated temperatures should be considered as possible in the formation of these deposits, the remobilization of Fe-oxides ± apatite along shear zones should be investigated.

# Chapter 10 Conclusion

This study represents the combination of modern techniques, like trace element compositions of magnetite and apatite by in-situ LA-ICP-MS, combined with practical techniques, such as outcrop observation, petrography, and whole rock geochemistry, in order to constrain the genesis of magmatic oxide deposits within the Lac St. Jean area. The results show that within amphibolite-granulite metamorphic terranes, trace element composition of magnetite has undergone significant change related to 1) the exsolution of ilmenite and Al-spinel from magnetite and 2) re-equilibration from Fe-Mg silicates. Concentrations of elements that were previously used to track fractionation in magnetite, such as Cr and V, are enriched during these processes therefore, reflecting equilibration conditions rather than crystallization conditions. Concentrations of Ni within magnetite has shown low degrees of change after exsolutions and/or re-equilibration with silicates when compared with the Ni concentrations from corresponding whole rock massive oxide. Therefore, Ni contents within magnetite could be a reliable indicator of magmatic evolution and show that Fe-Ti-V deposits crystallized from a more primitive magma than Fe-Ti-P deposits. The results also show that Ti-rich (approximately 21% $TiO_2$) magnetite is the sole liquidus Fe-oxide phase in all the Fe-Ti-V-P deposits/showins in the Lac St. Jean anorthosite. However, at Lac à l'Orignal, in the younger Vanel anorthosite, ilmenite is the initial liquidus Fe-oxide followed by magnetite. When magnetite is the sole liquidus Fe-oxide phase, modal proportions of ilmenite are limited by $fO_2$ of the magma because oxy-exsolution requires an $O_2$ source (possibly hydrous fluids) in order to proceed. It is also possible that Al-spinel formed through exsolution from Ti-magnetite and thus, the trace element composition of ilmenite and Al-spinel would not be indicative of primary magmatic conditions.

Apatite appears to be a more robust indicator of primary magmatic conditions than magnetite. Careful examination of its trace element composition indicates that the effects of sub-solidus processes which can alter the chemistry of apatite are minimal and therefore, trace element compositions obtained by in-situ LA-ICP-MS are considered to reflect igneous crystallization conditions. Chondrite-normalized REE patterns provided insight on three potential mantle sources involved in the genesis mineralization. Mineralization in the southern region was generated by a relatively similar degree partial melt to mineralization within the northern lobe. However, HREE fractionation appears to be significant within partial melts which formed mineralization within the northern lobe and negligible within partial melts which formed mineralization located in the southern region. The Lac à l'Orignal occurrence (hosted within the Pipmuacan AMCG suite) was generated from the lowest degree partial melt. Mineralization hosted within the Pipmuacan anorthosite suite also crystallized from a more oxidized magma than deposits hosted within the Lac St. Jean anorthosite based on Eu contents of apatite. This can also be correlated by the presence of hemo-ilmenite in the former relative to Ti-magnetite in the latter as the dominant Fe-oxide phase. Calculation of parental magmas using apatite was able to approximately reproduce REE distribution patterns from ferrodiorites (when present) and provided evidence for different mantle sources involved with the genesis of Fe-Ti-P deposits.

Many questions still remain regarding the methods in which Fe-oxides ± apatite became concentrated into massive ore bodies. The control of whole-rock chemistry on modality of Fe-oxides supports accumulation rather than immiscibility but, liquid immiscibility could be considered as an important process involved with the genesis of parental melts (i.e. ferrodiorites). The origin of massive oxide ± apatite bodies which crosscut the host lithologies still remains unknown and should be the subject of further examination. Although these complex questions remain unanswered upon conclusion of this project, valuable insights into the physiochemical conditions of the magmas involved with the genesis of Fe-Ti-V-P mineralization within the Lac St. Jean area were gained.